Old Electrical Log Interpretation (Pre-1958)

Douglas W. Hilchie

AAPG Methods in Exploration Series No. 15

Originally published by
Douglas W. Hilchie Inc., Boulder, Colorado

Reprinted edition published by
The American Association of Petroleum Geologists
Tulsa, Oklahoma, U.S.A.
Printed in the U.S.A.

Printed in the U.S.A.
Published May 2003

ISBN: 0-89181-666-6

AAPG Editor: John C. Lorenz
Geoscience Director: J. B. "Jack" Thomas
Special Publications Editor: Hazel Rowena Mills
Cover Design: Rusty Johnson
Production: SPI Publisher Services
Printing: The Covington Group, Kansas City, Missouri

This and other AAPG publications are available from:

The AAPG Bookstore
P.O. Box 979
Tulsa, OK 74101-0979
U.S.A.
Telephone: 1-918-584-2555 or 1-800-364-AAPG (U.S.A.)
Fax: 1-918-560-2652 or 1-800-898-2274 (U.S.A.)
www.aapg.org

Canadian Society of Petroleum Geologists
No. 160, 540 Fifth Avenue S.W.
Calgary, Alberta T2P 0M2
Canada
Telephone: 1-403-264-5610
Fax: 1-403-264-5898
www.cspg.org

Geological Society Publishing House
Unit 7, Brassmill Enterprise Centre
Brassmill Lane, Bath BA13JN
U.K.
Telephone: +44-1225-445046
Fax: +44-1225-442836
www.geolsoc.org.uk

Affiliated East-West Press Private Ltd.
G-1/16 Ansari Road, Darya Ganj
New Delhi 110 002
India
Telephone: +91-11-3279113
Fax: +91-11-3260538
E-mail: affiliat@nda.vsnl.in

I dedicate this book to Hubert Guyod who has contributed so much to the theory of old electrical logs and to Dr. R. G. Hamilton who has contributed much to the interpretation of old electrical logs. Both have helped me indirectly and directly with this book.

About the Author

Douglas (Doug) W. Hilchie has spent 50 years in the petroleum business. He earned a B.S. at the University of Oklahoma, an M.S. in petroleum engineering at the University of Texas, and a Ph.D. in engineering sciences at the University of Oklahoma. His professional employment has been with Schlumberger, Cities Service, Mobil, Dresser Atlas, the University of Oklahoma, Montana School of Mines, and Colorado School of Mines, and as a consultant with Douglas W. Hilchie Inc.

Hilchie has published five books on logging and log interpretation, five in-house log-interpretation manuals for major oil companies, and numerous technical papers. He has taught courses in log interpretation all over the world, has designed a log-interpretation course for home study, and has worked as a consultant in log interpretation. He has been a member of the American Association of Petroleum Geologists, the Society of Professional Well Log Analysts, the Society of Petroleum Engineers, and the Society of Exploration Geophysicists.

Table of Contents

Foreword

Douglas W. Hilchie originally published this book in 1979 to fill a need at that time; there was no book dedicated to analysis and interpretation of wireline logs run in wells drilled prior to 1958. Today, many oil and gas fields in the world that were drilled before 1958 are still producing. Current operators of those wells have lost the opportunity to learn proven methods of interpretation of log suites and logging-tool technology of that vintage.

This book contains charts for rapid log interpretation generated by companies that have long been out of business. In reprinting Dr. Hilchie's benchmark text, we include the best-quality figures available from the originals. We believe in all cases that the quality is sufficient for you to use them in evaluating your wells.

Although Dr. Hilchie has been semiretired since 2001, he continues to reflect on the challenges of "getting more information out of old logging suites." Perhaps in his more relaxed years, he will write his next definitive book on interpretation of old electrical logs 1958–1980!

We owe our gratitude to Dr. Hilchie for his efforts in producing this classic publication, as well as for his years of dedicated mentoring and instruction in the industry. Universities, logging companies, and integrated energy companies have claimed him as an employee and contributor, but the industry claims him as a dedicated teacher. To Dr. Hilchie, we extend our deepest appreciation. May this book, his contribution, help you in your pursuit of the hidden target — oil and gas!

AAPG
2003

Preface to the Reprinted Edition

Old (pre-1958) electrical logs comprise as much as 40% of the well-log data in oil and gas company files. Books and technical papers related to these unique well logs are very limited and, in most cases, almost unobtainable in today's world. With fewer log analysts and petrophysicists understanding the analysis of these logs, it has become more important that geologists using this data obtain a better knowledge. Today, even though interpretation of well-log data is at best semiquantitative, experience has shown that in 70–80% of definitive interpretations, recompletion of these wells produces commercial hydrocarbon production.

Geologists should understand that interpretation of these logs has improved significantly in the last 50 years. It should also be noted that economics and production techniques have changed significantly in that time period. What may have been overlooked or ignored in past times is now economically viable.

Douglas W. Hilchie
2003

Introduction

This book covers the interpretation of old electrical logs. Common usage of the term electrical log or (from even older times) Schlumbergers often refers to all measurements made on the end of a wireline. In this book, the term electrical log (or the abbreviation ES) refers to logs which are combinations of SP's, normals and laterals. The major portion of this book is dedicated to electrical logs although other logs of this time span (1927 – 1958) such as Micrologs, laterologs, microlaterologs, limestone logs and the old gamma ray neutron logs are covered briefly. The name used for these logs or the equivalents, such as Microlog which is a Schlumberger trademark, are the most commonly used. In many cases, the usage is so common that the term is essentially a generic name. The final chapter of this book introduces the use of old electrical logs in conjunction with modern through casing pulsed neutron logs (Neutron Lifetime and TDT logs). In short this book picks up the pieces not covered by modern well log interpretation books.

The first electrical log was run by the Schlumbergers in 1927. It was a series of stationary measurements. The electrical log progressed with time until in about 1938 it reached a stage of maturity such that changes after that were refinements and not giant steps. In 1956 the induction log started to replace the electrical log as the prime resistivity measurement. By 1958 that replacement was essentially complete. The role of the electrical log and the newer induction log were the same, only the induction log did it better. Shortly after the introduction of the induction log came the acoustic velocity log, the first successful porosity log, and modern log interpretation was born. The problem is that in mature petroleum provinces such as the Gulf Coast, Venezuela, Oklahoma, etc., there were more electrical logs run than any other kind. Also, with the present emphasis on enhanced oil recovery from old oil fields, the geologists and petroleum engineers must deal with old electrical logs on a day-to-day basis.

The interpretation of old electrical logs is usually based on qualitative porosity data. An old adage in log interpretation is that if you don't know porosity, you cannot determine if hydrocarbons are present or "no porosity - no interpretation". In old electrical log interpretation, we try to violate this adage. Porosity logs came after the old electrical log was at its prime. Accordingly, we will use every possible way available to obtain some indication of porosity from these old logs. We will also pay close attention to detail so that any hydrocarbon indication on the curves will be observed so that we can identify the existence of hydrocarbons in a zone and then try to determine how much (water saturation).

This book assumes some knowledge of log interpretation in that such common items as the SP are only briefly covered. If you are a novice, it is recommended that you read some modern log interpretation book* before going too deeply into old electrical log interpretation.

*e.g., APPLIED OPENHOLE LOG INTERPRETATION by D. W. Hilchie (1978)

1

Hilchie, D. W., Resistivity and SP interpretation fundamentals review, Old electrical log interpretation, p. 1–8.

Resistivity and SP Interpretation Fundamentals Review

Most reservoir type formations we deal with are assumed to be clean (i.e., the shale influence is minimal, usually less than 15%) and we use the "Archie" equation to determine water saturation. In this text I have purposely used old symbols so that the reader can, if he wants to, go back and read the old pertinent literature with few symbol problems. Archie's equation is:

$$Sw = \sqrt{\frac{Ro}{Rt}} = \sqrt{\frac{F\,Rw}{Rt}} \qquad (1\text{-}1)$$

where

Sw is water saturation with a normal range of 10 to 100%

Ro is the resistivity of a reservoir rock with Sw = 100%

Rt is the true resistivity of the reservoir rock at any Sw

Rw is the formation water resistivity

F is the formation resistivity factor which is = $\frac{Ro}{Rw}$

which is related to porosity by

$$F = a\,\emptyset^{-m} \qquad (1\text{-}2)$$

where a and m are empirical constants and Ø is porosity

The most commonly used versions of equation 1-2 are:

For Sandstones,

$$F = .62\,\emptyset^{-2.15} \qquad (1\text{-}3)$$

which is the Humble relationship, and

$$F = \emptyset^{-2} \qquad (1\text{-}4)$$

which is used for nongranular reservoir rocks such as most carbonates. Of course, in those cases where better values of m are known for a particular formation, they should be used.

A nomographic solution to equations 1-1, 1-3 and 1-4 is shown in Figure 1-1. The F – Ø relationships have been circumvented by entering porosity directly. On the nomograph Rw and porosity (Ø) are used to calculate Ro, Ro and Rt are then used to calculate Sw. An example is worked below the nomograph. If porosity is constant and a water bearing zone exists, you may read Ro directly off the log and use it with Rt to obtain Sw.

The quality of Sw obtained is, of course, controlled by the quality (or correctness) of the numbers you put

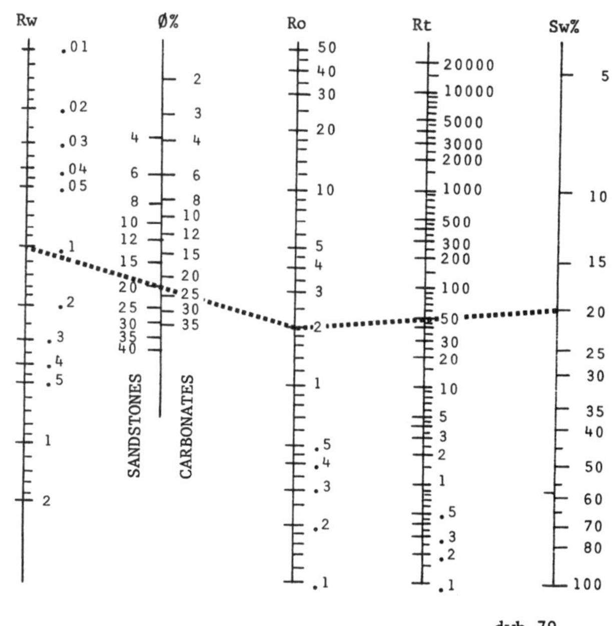

Example Rw = .1 Sandstone Ø = 20%
calculated Ro = 2
Rt = 50
calculated Sw = 20%

Figure 1-1. Nomograph Solution to Equation 1-1

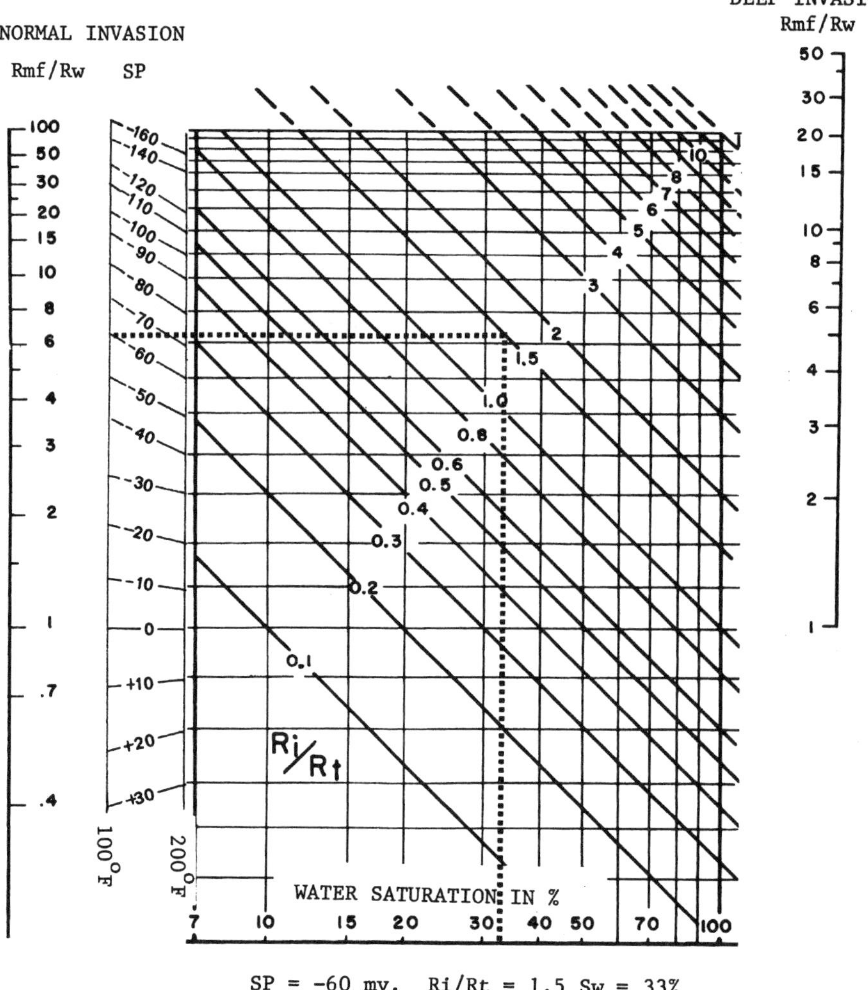

Figure 1-2. Rocky Mountain Method for Sw (after Tixier, Oil and Gas Journal June 23, 1949)

in the nomograph or equation. The greatest problem in old electrical log interpretation is the determination of a good porosity. The other problems are the determination of Rw and the determination of Rt in thin resistive beds. In short, we have problems with all three variables in equation 1-1. For this reason, the water saturations we calculate are (most) often just estimates.

It is for the above reasons that methods such as the Rocky Mountain method shown in Figure 1-2 were developed. The Rocky Mountain method is usable in hard rock areas where normal or deep invasion of the mud filtrate occurs. This system does not work well in most parts of the Gulf Coast where invasion in minimal. The method uses the short normal (Ri) as a porosity indicator and the SP as a formation water resistivity indicator. Rt for this chart should be obtained from the lateral curve. Two empirical corrections are supplied, one for normal invasion and the other for deep invasion. The correction applied assumes a mixing of the mud filtrate and formation water in the invaded zone.

Figure 1-3 is a derrick floor interpretation chart derived from Figure 1-2. Figure 1-3 shows the relative separation between the Ri (short normal) and Rt (lateral) curves for a water zone versus a hydrocarbon bearing zone of 40% water saturation. This chart may be used qualitatively to distinguish hydrocarbon from water bearing zones.

If invasion is shallow, the Rocky Mountain method may not be used and porosity must be obtained from some other source. If porosity is constant and water zones are available, Figure 1-1 can be used directly by in-putting Ro and Rt from adjacent water and hydrocarbon bearing zones.

In shaly formations it was common to compensate for the shale by varying the water saturation exponent from 2 (Sw^n) and the porosity exponent together. One set of these compensations used by Dr. R. G. Hamilton is:

Vsh %	*n*	*m*
10	1.8	1.8
20	1.7	1.7
25	1.6	1.6
30	1.5	1.5

Vsh is volume of shale.

These obviously had no theoretical basis but produced reasonable water saturations in shaly formations.

SPONTANEOUS (SELF) POTENTIAL (SP)

The SP is a naturally occurring potential in the wellbore. The potential is developed opposite permeable beds when an electrochemical cell, composed of shale, fresh water and salty water is active. The cell-

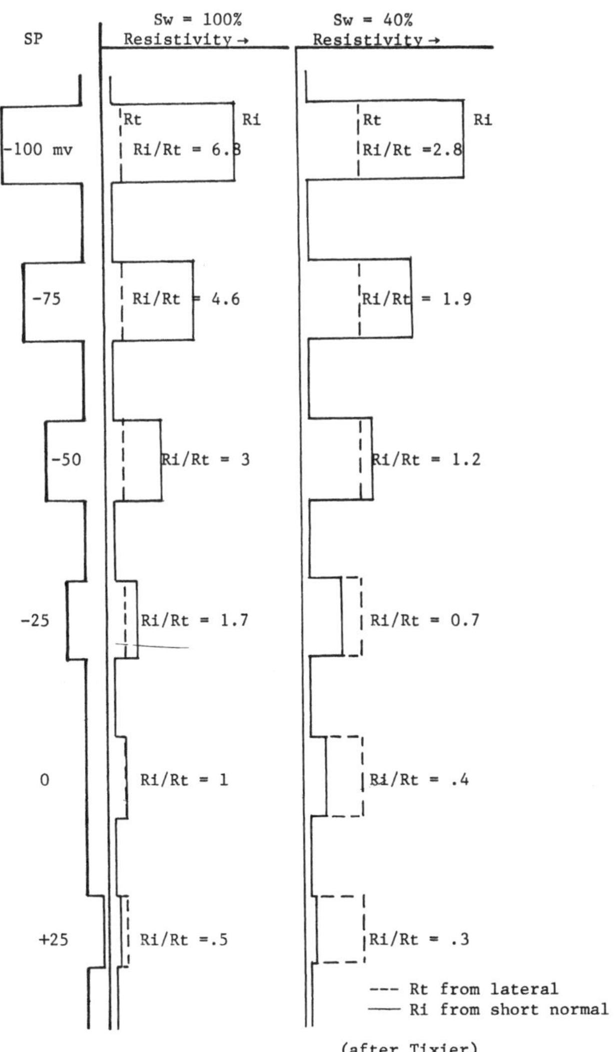

Figure 1-3. Derrick Floor Interpretation Chart

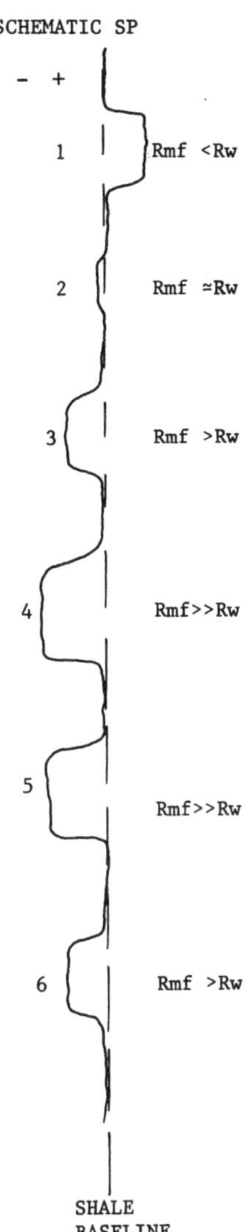

Figure 1-4.

developed voltage is primarily an indicator of permeability. The amount of voltage generated is due to the contrast of the borehole water (mud filtrate) and the formation water salinity (or resistivity). Figure 1-4 is a schematic SP. The shale baseline (dashed in Figure 1-4) is the reference for all observations. If the SP shift is to the right (+) of the shale baseline, the formation water has a higher resistivity (Rw) than the mud filtrate (Rmf). That is the mud is saltier than the formation water. Zone 1 on Figure 1-4. If the Rmf is greater than the Rw, a negative (–) SP develops. Zone 3. The larger the contrast between Rmf and Rw, the larger the SP deflection. The SP can thus be used to qualitatively and quantitatively estimate Rw as long as the mud is homogeneous in the borehole, which is the norm. If the SP has about the same deflection from the shale baseline for several zones (Figure 1-4

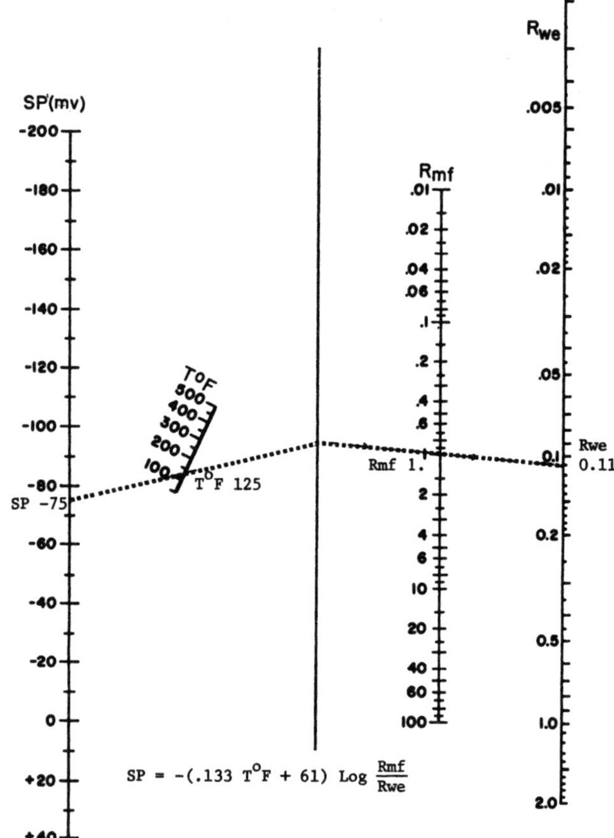

Figure 1-5. First Part of Rw from SP Nomograph (adapted for fresh muds Rm ⩾ .1 ohm m)

zones 4 and 5), it can be assumed that Rw (the formation water resistivity) is constant. On Figure 1-4, zones 1-6 are permeable.

The SP can be used to estimate the formation water resistivity (Rw) that is needed to determine water saturation in equation 1-1 (or Figure 1-1). The determination of Rw from the SP is at minimum a two-step process. The first step requires the solution of the fresh mud version of the SP equation:

$$SP = -(.133\ T°F + 61)\ \log \frac{Rmf}{Rwe} \qquad (1\text{-}5)$$

using Figure 1-5 (or the equation) to obtain Rwe. Rwe is then transformed to Rw using Figure 1-6. The SP used is the millivolts – or + from the shale baseline, the temperature in the borehole at the particular formation depth at the time of logging, and the mud filtrate resistivity corrected to the temperature of the borehole opposite the formation of interest. The procedure described in this book must be revised if the mud is gyp based. Other more complete texts should be consulted for this process.

Rw FROM THE SP

A quick determination of the Rw from the SP (this Rw should always be considered an estimate) may

Figure 1-6. Second Part of Rw from SP Determination — Rwe to Rw transformation

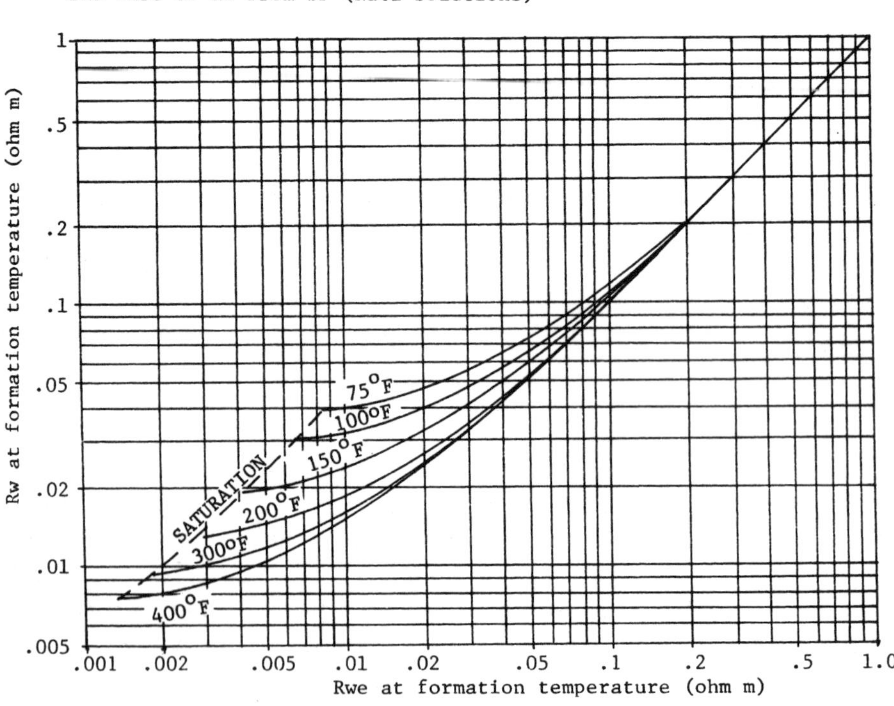

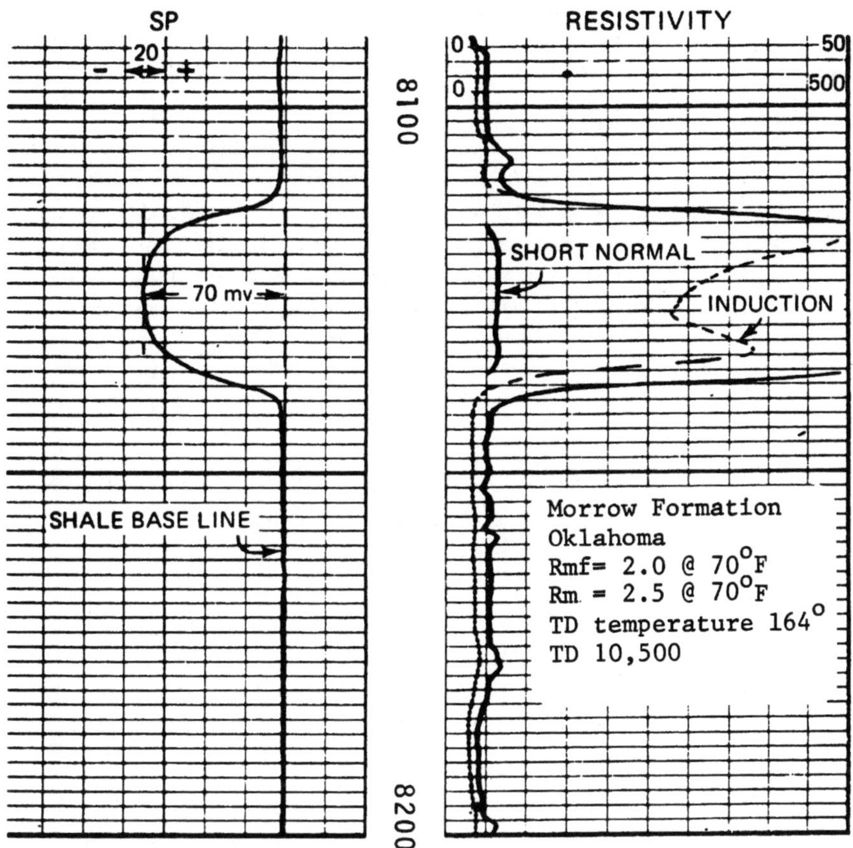

Figure 1-7. Rw from SP example log.

1. Establish shale baseline as shown
2. Determine temperature at 8100 feet using Figure 1-8
   BHT = 164° at 10,500 feet (TD)
   Surface temperature for Oklahoma is 60°F
   From chart Figure 1-8 Tf at 8100 feet is 140°F
3. Correct Rmf to 140° using chart Figure 1-9
   Rmf = 1.0 @ 140°
4. Obtain SP reading from curve - - -70 mv
5. Using SP reading, 140° and Rmf = 1.0 from above steps go to Figure 1-5
   Rwe = .13 ohm m
6. Using Figure 1-6 correct Rwe to Rw
   Rw = .13 ohm m (in this case there is no correction)

be done as described in the previous section. In this section we will work an example to remind you of the process. Figure 1-7 is a log over the Morrow sandstone.

REVIEW OF SP CURVE SHAPES

A good understanding of the SP for old electrical log interpretation requires a knowledge of how the SP responds under given conditions. The previous section should have reinforced the idea that the amplitude of the SP deflection in permeable zones is controlled by the mud and formation water salinities. If the formation water is the same and you drill with saltier mud, the SP will have smaller anomalies opposite the permeable beds.

Figure 1-10 is a review of SP curve shapes. Figure 1-10a shows a thick permeable bed with the SP anomaly opposite the permeable bed being flat. If the bed is thin, Figure 1-10b is what you will get, a pointed SP. In all cases the bed boundary is at the inflection point as shown on the figures. If the shale and permeable bed have the same resistivity the inflection point will be half way between the shale and permeable amplitudes of the SP. If the shale resistivity (Rsh) is much larger than the permeable bed resistivity (Rsd in this case), the inflection point will move towards the permeable bed line (Figure 1-10c). If the formation that is permeable has a much larger resistivity than the shale, the inflection point will move towards the shale baseline (Figure 1-10d) and the SP will be more rounded opposite the permeable zone. Figure 1-10e shows the result of increased amounts of shale on a

Figure 1-8. Estimation of formation temperature

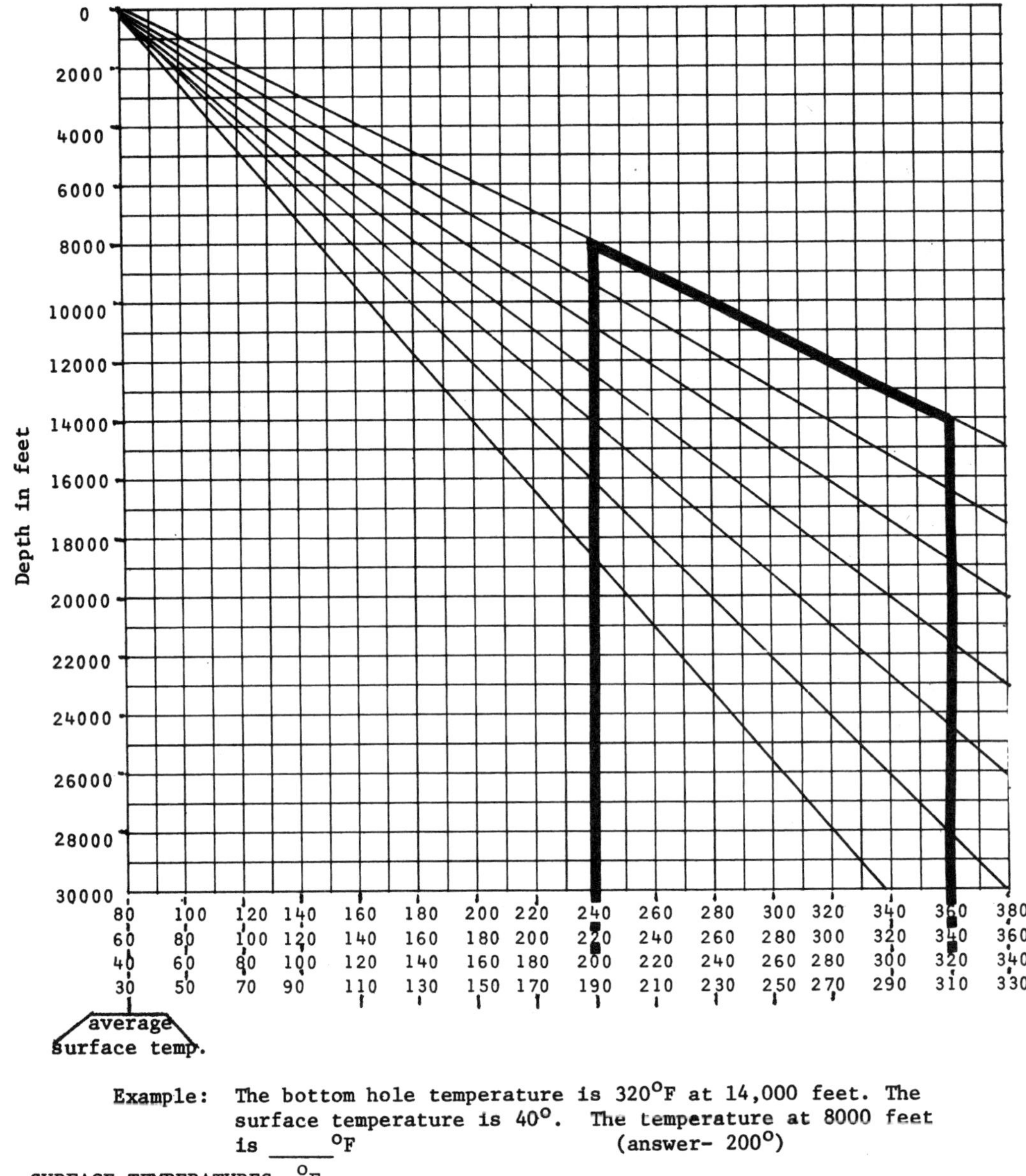

Example: The bottom hole temperature is 320°F at 14,000 feet. The surface temperature is 40°. The temperature at 8000 feet is _____ °F (answer- 200°)

SURFACE TEMPERATURES °F

Gulf Coast	- 70°	Wyoming & Montana	- 40°
Oklahoma	- 60°	Alberta	- 35°
Colorado	- 50	California	- 60°

water bearing permeable zone. The larger the shale content the lower the SP amplitude until at 100% shale the curve will lie on the shale baseline. Figure 1-10f shows a permeable bed with many thin permeable and shale stringers interbedded.

If a formation is shaly and contains hydrocarbons, the SP is reduced by an undeterminable amount. This phenomenon is called hydrocarbon suppression. The amount of suppression does not seem to be controlled by the amount of shale or any other of the variables we observe. In clean formations, the suppression is zero or essentially zero. Since most formations contain some shale, hydrocarbon suppression is common. Unfortunately, we often cannot tell if the SP reduction is due to shale or hydrocarbons. Figures 1-10g & h show hydrocarbon suppression. Figure 1-10h shows a common looking SP where there is a water bearing zone under a hydrocarbon bearing zone.

Figure 1-11 reviews the SP curve shapes in carbonates where there are tite (impermeable) zones. The key to understanding these shapes is that the movement of the SP requires an Electrochemical cell. This means that for the SP to move left (in a fresh mud situation), the formation must be permeable. For the SP to move

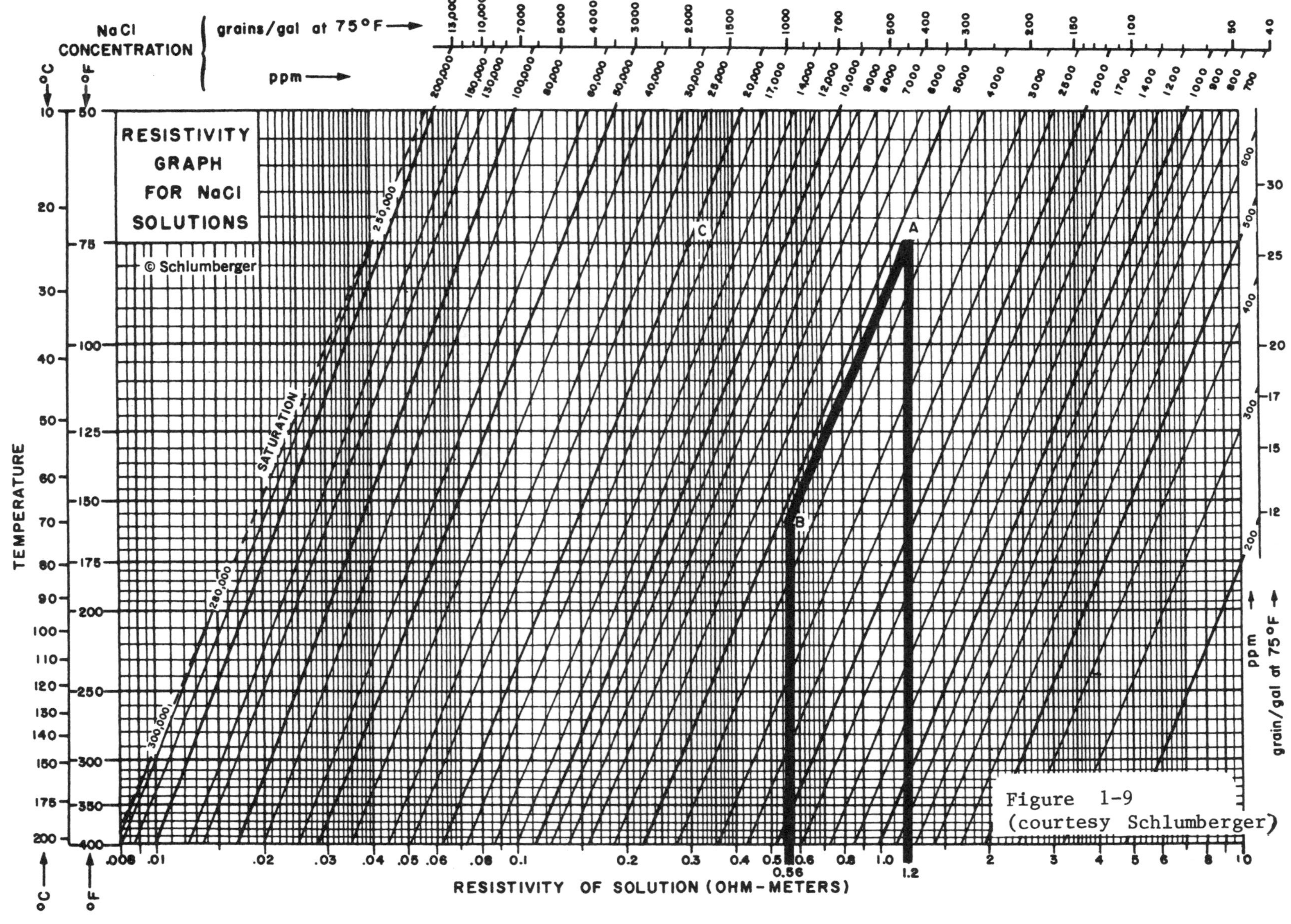

Example: R_m is 1.2 at 75°F (point A on chart). Follow trend of slanting lines (constant salinities) to find R_m at other temperatures; for example, at Formation Temperature (FT) = 160°F (point B) read $R_m = 0.56$. The conversion shown in this chart is approximated by the Arps formula: $R_{FT} = R_{75°} \times (75° + 7)/(FT \text{ (in °F)} + 7)$.

Figure 1-9.

back to the shale baseline requires a shale. Tite zones restrict the SP current flow to the wellbore and result in straight angular lines. Figures 1-11a & b show a permeable zone on top and bottom of a thick nonshale.

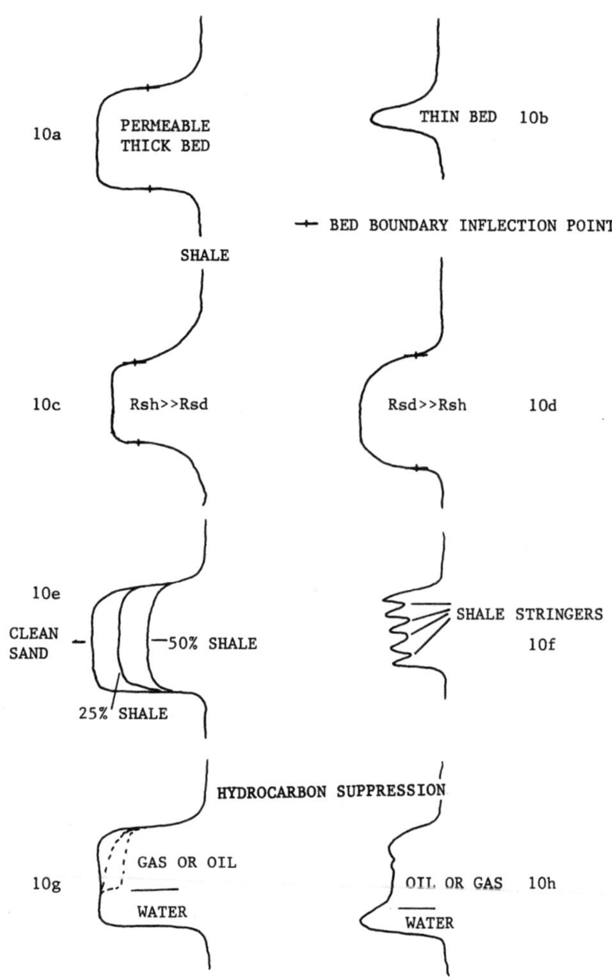

Figure 1-10. A Review of SP Curve Shapes

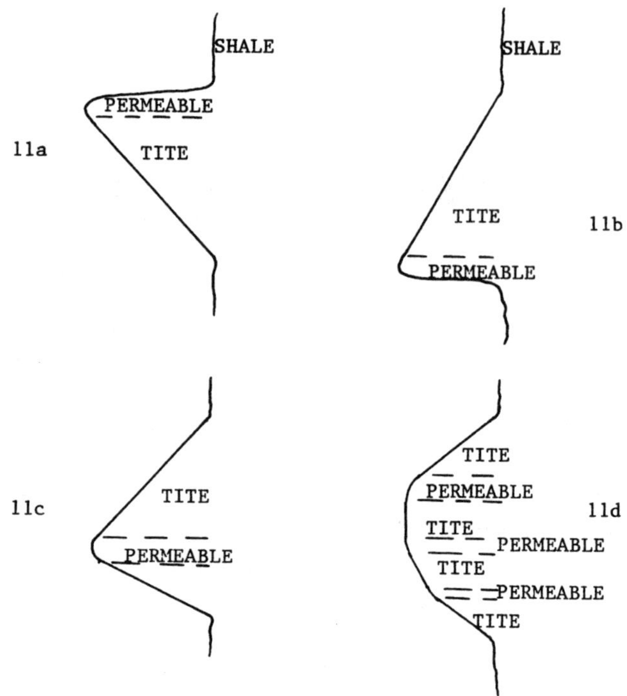

Figure 1-11. REVIEW OF SP CURVE SHAPES IN CARBONATES

Figure 1-11c shows a permeable zone with tite stringers above and below. Figure 1-11d shows a complex combination of tite and permeable stringers. In carbonates that are thick it is almost impossible to determine the location of permeable beds with the SP. This is partially due to the fact that the shale baseline shifts and after a few hundred feet it is often impossible to know where the shale baseline is.

The SP is a very complex curve and to really determine its strengths and weaknesses requires much more study than we have devoted to it here. I would suggest referring to another text for more details.

Hilchie, D. W., The lateral curve, Old electrical log interpretation, p. 9–20.

2

The Lateral Curve

The lateral was the first curve run by Schlumberger in 1927. It was not until 1932 that any other resistivity curve was run. The more modern electrical logs are combinations of lateral and normal curves with different depths of investigations. The lateral is the best source of Rt from the electrical log when the beds are sufficiently thick.

To develop a better "feel" for electrode resistivity measuring systems I will introduce the mud cup.

THE MUD CUP

The mud cup may be used to measure the resistivity of any aqueous solution but was designed to measure mud resistivity. It is also quite similar to the way resistivity measurements are made in the borehole. The cup itself is some insulating material like bakelite or rubber. Electrical current is passed through the liquid in the cup by an electrode placed at each end. The electrical current used in most electrical logs was a pulsated DC (or a low frequency square wave). The pulsating was at 10 cycles per second. The current came from batteries and was chopped up with a pulsator. It is for this reason that all electrical theory discussed will use DC (direct current) theory.

From electrical theory:

$$r = R\frac{L}{A} \qquad (2\text{-}1)$$

where

r is the electrical resistance in ohms

R is the electrical resistivity in ohm meters

L is the length in meters

A is the cross sectional area in square meters which the current flows through

$$V = I\,r \qquad (2\text{-}2)$$

which is ohms law

where

I is currents in amperes and

V is voltage or potential in volts

For the mud cup (Figure 2-1) the combining of equations 2-1 and 2-2 becomes:

$$V = I\,R\,\frac{L}{A} \qquad (2\text{-}3)$$

which simplifies to:

$$R = V\,K'$$

where K′ is a constant which includes the area A through which the currents flows, L the length of mud through which the current flows and which is designated as the distance MN. M and N are the electrodes across which the voltage is measured. The actual material being measured is between M and N. Also in K′ is I, the current which is constant.

A and B on the mud cup are the electrodes through which current is passed through the sample.

Equation 2-4 thus says that for any system with fixed electrodes (A, B, M & N) and a constant current, the change in voltage drop between M and N is directly related to the resistivity.

THE LATERAL CURVE

The lateral electrode array is shown in Figure 2-2. The current flows from electrode A to B; the latter being located at the surface or a very long distance away. The current flows radially if the material surrounding the system is isotropic and homogeneous and you stay relatively close to electrode A. The two spheres at distances L and L + dL from A are surfaces on constant potential (voltage). The formation or material being measured is between these two spheres as the voltage drop is being measured from M to N. The mathematics relating voltage to resistivity are presented as part of Figure 2-2.

The electrode spacing on the lateral is usually referred to as A0. 0 is the mid-point between M and N and is the depth reference point (zero for depth or the point where the resistivity is recorded on the log). The lateral is defined as a three electrode curve where $AM \ll MN$.

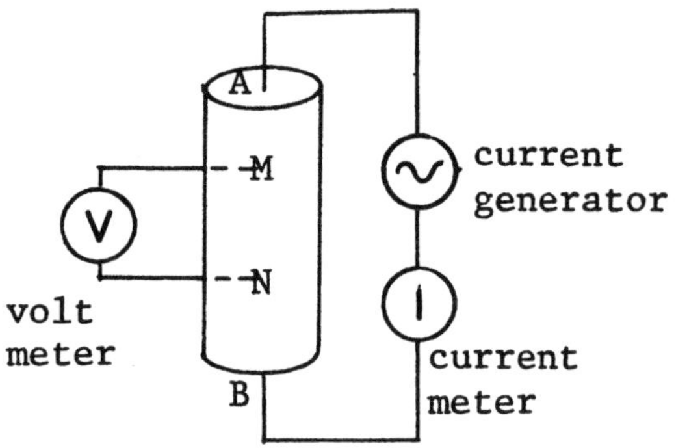

Figure 2-1. The Mud Cup

The electrode spacing, A0, determines the depth of investigation of the lateral curve. Usually, the depth of investigation is considered to be the spacing A0. A0 also affects the curve shapes by making decay zones, dead zones and reflection points closer or wider apart. The most common value of A0 was 18′8″ although 5, 6, 7.5, 9, 13, 15, 16, and 24 foot spacings were run. The more common spacings as a function of time and service company are listed in Chapter 4.

The lateral curve is generally used to estimate Rt if the bed in question is thick enough.

The terms resistive and conductive beds are used in this text as they are in many others. Although they are not precisely accurate, I find them necessary. A resistive bed is one that has a higher resistivity than the adjacent beds. A conductive bed is one that has a lower resistivity than the adjacent beds. The adjacents beds, usually referred to with the symbol Rs, may be shales or beds of lower or higher resistivity next to the bed in question all of which may have the same composition; e.g., sandstone, limestone, etc.

LATERAL CURVE SHAPES

At the very beginning, the lateral curve was run for four years before it was observed that the curve was asymmetric. That is, in a bed of constant resistivity the common lateral curves show peaks at the bottom and low readings at the top. Figure 2-3 shows the theoretical curve shapes of a lateral curve for a thick resistive bed. The dashed curve is for no borehole present and the solid curve is for a borehole present. The top of the bed shows a notch of lower resistivity than the adjacent bed. Below this, for one electrode spacing, A0 is the "decay zone" the value of which is controlled by the adjacent and bed proper resistivities. The curve then increases (in resistivity) until a plateau is reached (in very thick beds) after which the resistivity increases to a maximum at the bottom of the bed. Below the bed the resistivity moves towards the lower adjacent bed resistivity for about one A0 distance.

The sharpness of the peaks and notches on the curve are a function of the borehole size and the resistivity of the mud. An increase in borehole diameter and/or a decrease in mud resistivity will round the curve as shown in Figure 2-4. Curves, such as Figure 2-4, were obtained by Hubert Guyod on an analog model which exactly simulated the real conditions

Figure 2-2. The Lateral Electrode Array

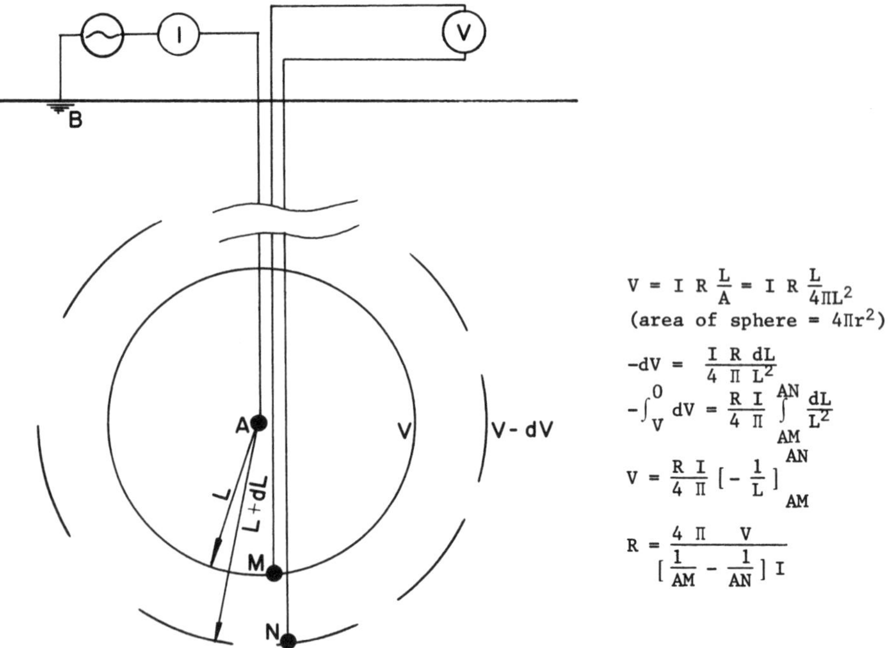

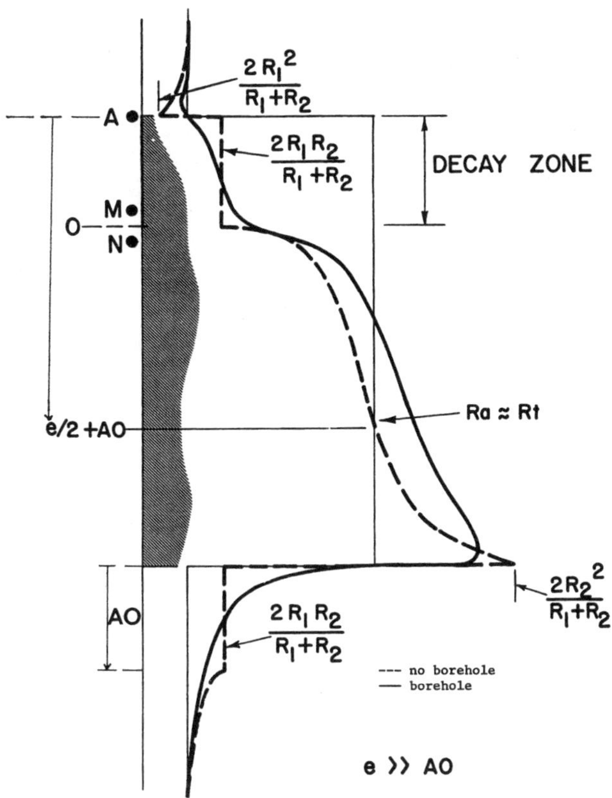

Figure 2-3. Lateral Curve Shape in a Thick Resistive Bed (after L.G.Chombart 1949)

As a thick resistive bed is reduced in thickness, the curve shapes stay much the same, only it appears that the reduction in thickness is being taken out of the center of the bed. The peak value at the bottom of the bed is still higher than Rt until the bed thickness reaches 1.3 A0 (24 feet for an 18′8″ lateral). Figure 2-4 is about this thickness. As the thick resistive bed thins, the curve starts looking like the d = 0 curve on Figure 2-4. The peak that starts to occur A0 below the bed boundary is called a "reflection peak". Figure 2-5 shows the curve shape for a resistive bed twice A0 in thickness.

In this text modern symbols are not used. The symbols used in the 1950's and before are used so that of logging. Hubert Guyod has graciously permitted our use of these figures.

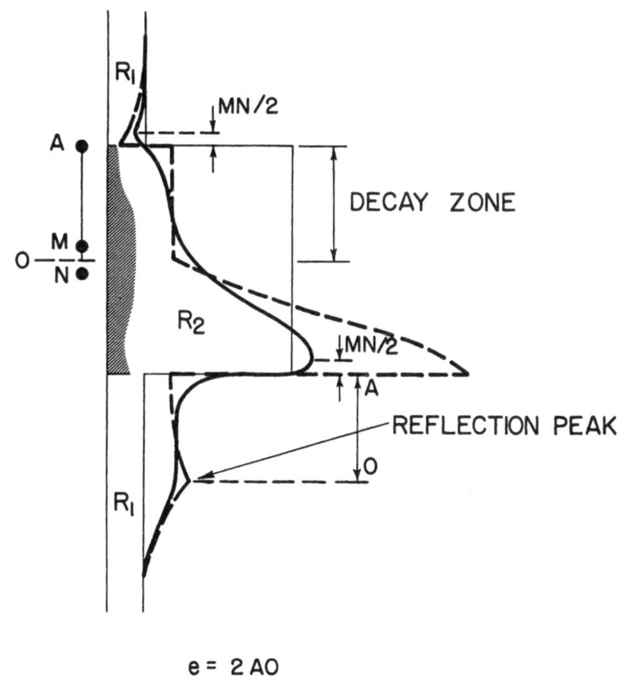

Figure 2-5. Lateral Curve Shape in a Resistive Bed (after L.G.Chombart)

From a practical point of view, the peak at the bottom and the notch at the top of the bed (if you can find it) may be used to designate the bottom and top of the bed. In fact, they are displaced upward by one-half the MN distance, which for the 18′8″ lateral was 32″ ÷ 2 = 16 inches.

Rt is obtained in thick resistive beds on the curve one A0 distance below the center of the bed (Figure 2-3). Ra is referred to as the apparent resistivity from the curve.

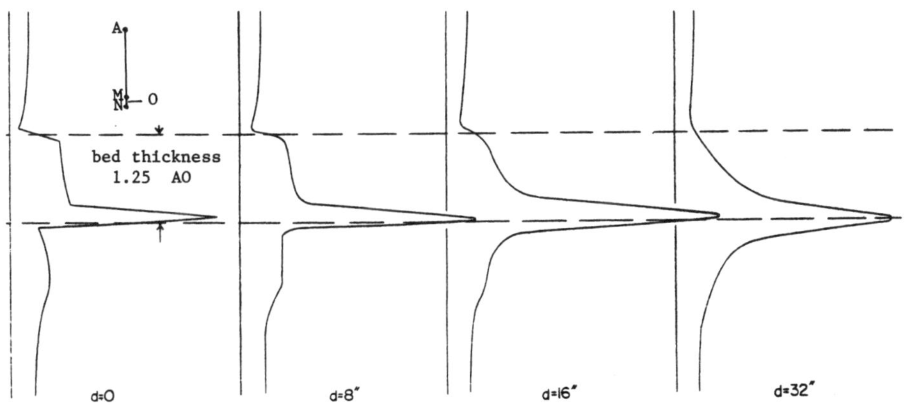

Figure 2-4. A Comparison of Lateral Curves in Four Borehole Diameters (courtesy Hubert Guyod)

any additional reading you do will not involve relearning the symbols.

As the bed thickness approaches the A0 spacing, the resistive bed seems to disappear. This is called the critical thickness.

A resistive bed thinner than A0 is shown in Figure 2-6. The resistive bed shows up on depth on the curve. A0 below the bottom of the bed the reflection peak shows up. Between the reflection peak and the bed is the "dead" zone. In the dead zone the lateral curve is of little to no value. Some resistive anomalies will show up but not many.

For a thin resistive bed such as in Figure 2-6, the peak resistivity value at the bottom of the bed and the value read in the dead zone are related. The higher the resistivity in the bed, the lower the dead zone resistivity. If the Rt must be estimated in this type of thin bed

$$Rt > Rs \frac{(\text{Ra in the bed})}{\text{Ra in the dead zone}} \quad (2\text{-}4)$$

The response of the lateral in thin resistive beds is useful in determining the A0 spacing when it is unknown, as it often is in pre-1950 logs. You must know the A0 lateral spacing to interpret the log. The curve shape is a function of the bed thickness and A0 distance.

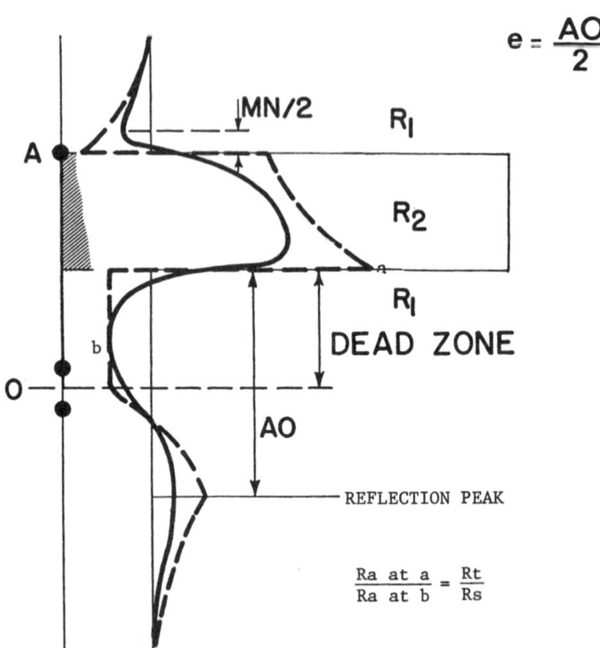

Figure 2-6. Lateral Curve Shape in a Resistive Bed Thinner than A0

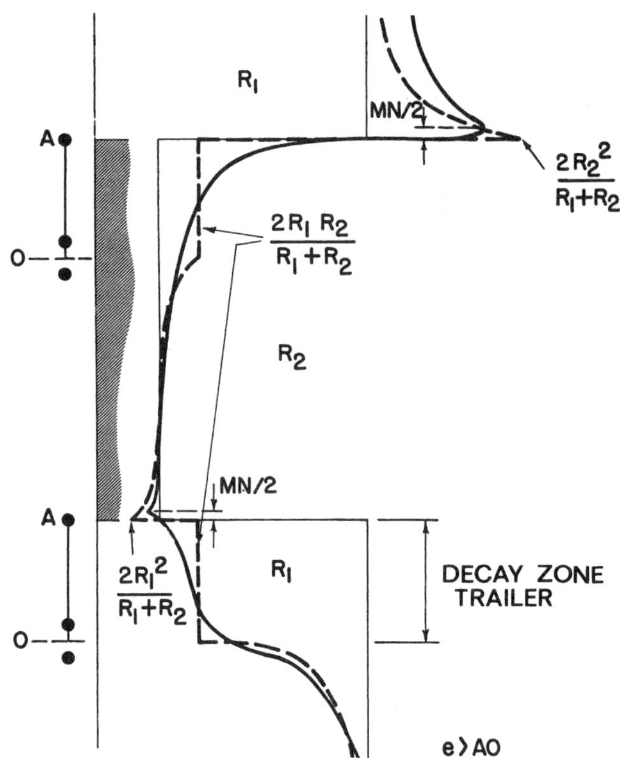

Figure 2-7. Lateral Curve Shape for a Thick Conductive Bed (after L.G.Chombart)

Conductive beds

Figure 2-7 shows the lateral curve shape for a conductive bed (a bed with a resistivity less than the adjacent beds). The apparent resistivity in a conductive bed is closer to Rt than for a resistive bed. The top of the bed is the resistivity peak. The curve takes about A0 distance to reach Ra = Rt and then stays low for the rest of the bed. There is a notch at the bottom of the bed and the resistivity curve stays low for one A0 distance. This is the decay zone which I will refer to as the trailer for conductive beds. The resistivity reading in the trailer is contolled by the resistivity of the conductive bed above and the resistivity of the bed below. It is thus an indicator of the resistivity of the bed above. The trailer remains as the conductive bed becomes thinner and thinner.

Figures 2-8 and 2-9 show lateral curves for thin conductive beds. Note that the apparent resistivity from the lateral (dashed curve) is close to Rt until the bed becomes very thin. For thin conductive beds, read the lowest apparent resistivity (Ra) from the lateral curve. This usually occurs in the center of the bed or the lower half of the conductive bed.

Figure 2-10 shows the influence of the borehole on the lateral curve in a conductive bed. As with the resistive bed, the curve becomes more rounded as the

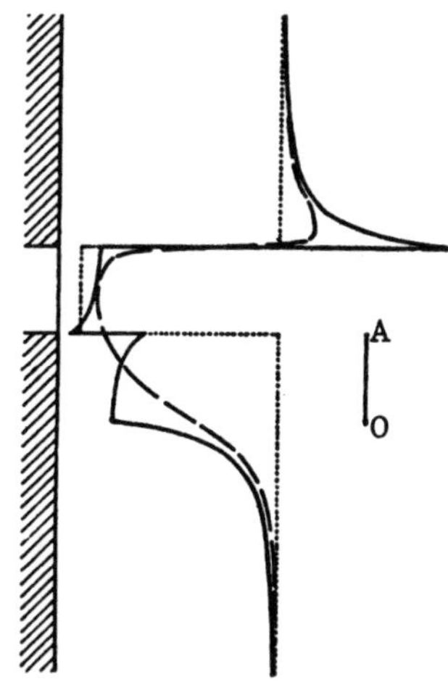

Figure 2-8. Lateral Curve for a Thin Conductive Bed (courtesy Welex)

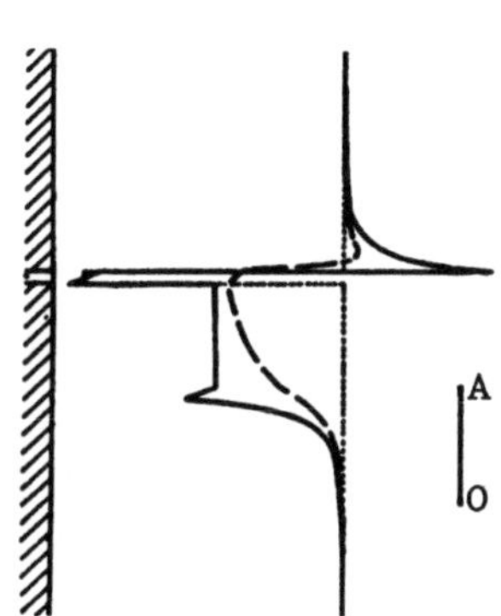

Figure 2-9. Lateral Curve for a Thin Conductive Bed (courtesy Welex)

borehole size increases. A reduction in mud resistivity in the borehole will have much the same effect.

Caution must be used if you run into one of the early Schlumberger or Lane Wells electrical logs which use an inverted lateral. In these cases the curve responses will be inverted or if you wish upside-down. This will be true of both the conductive and resistive beds. Figure 2-11 shows a comparison between a regular and inverted lateral curve in a resistive bed.

$$Ra \simeq Rt \ (Rat)$$

Most people, when dealing with the lateral, use the best apparent resistivity (Ra) reading from the curve as Rt and do not correct for borehole, adjacent bed

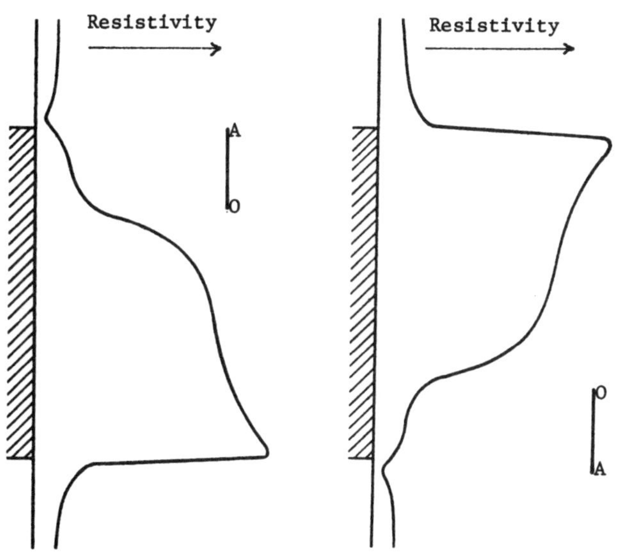

Figure 2-11. The Regular and Inverted Lateral in a Resistive Bed

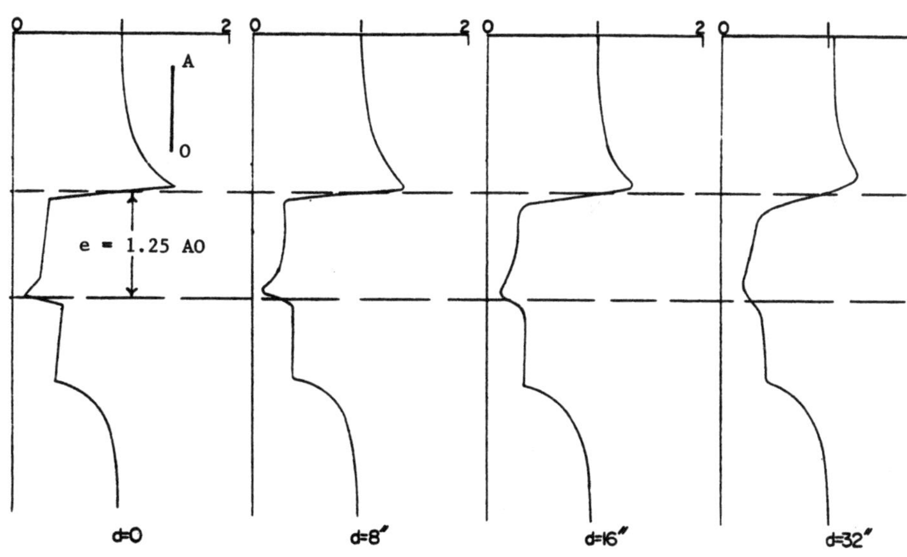

Figure 2-10. Influence of the Borehole on the Lateral in Conductive Beds (courtesy Hubert Guyod)

and invasion effects. (These will be covered later in this chapter or in the case of invasion, later in the book under departure curves.)

Rules

Conductive beds - read the lowest resistivity in the lower half of the bed.

Resistive beds - Follow the rules shown in Figure 2-12.

Mid-Point Rule — Find the point half way between the top and bottom of the bed and then go down one A0 distance and read the resistivity directly off the curve. This is Ra.

2/3 Rule — From the top of the bed go down one A0 distance and read resistivity, at peak value of resistivity read, Ra is 2/3 between these two values.

Point Rule — When the bed is 1.3 times A0 (for 18′8″ lateral bed must be $24^{1}/_{4}$ feet thick), Ra is the peak value.

Thin Rule — When the bed is thinner than A0 distance a very approximate value of Ra may be determined. (see Figure 2-12 for equation)

The transition thicknesses between these rules can be very difficult. Figure 2-13 was developed to aid in picking values. On Figure 2-13 all distances are either

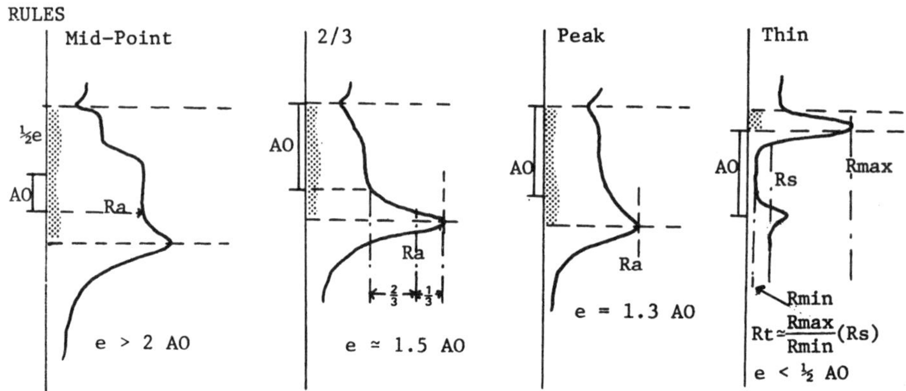

Figure 2-12. Ra approximates Rt Rules for the Lateral in Resistive Beds

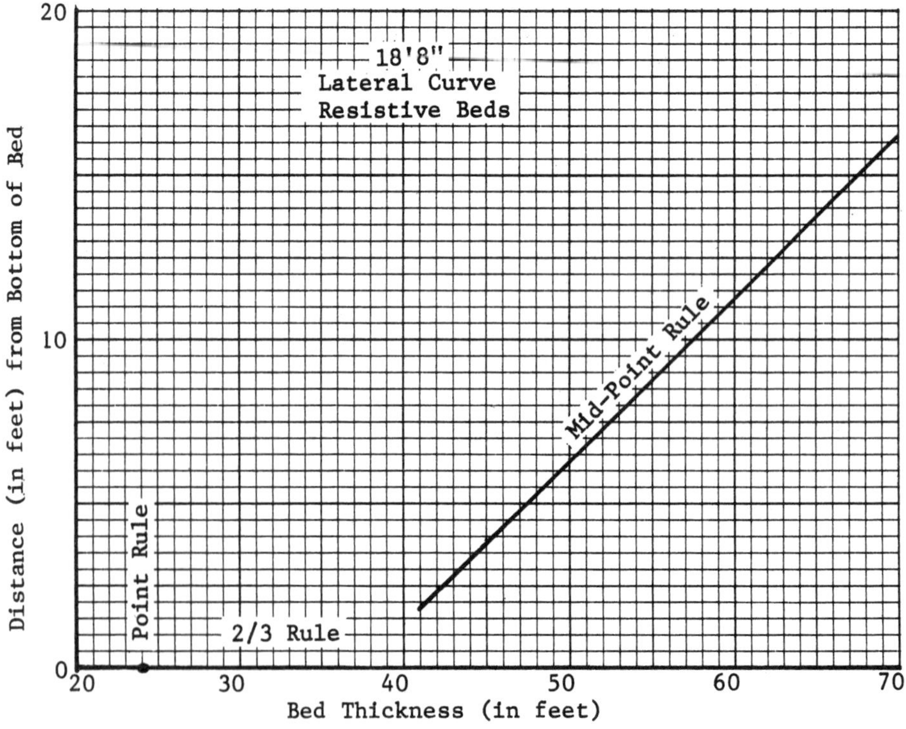

Figure 2-13. 18′8″ Lateral Resistive Bed Rule Graph

relative to A0 or are picked from the maximum peak at the bottom of the resistive bed.

Note: If an inverted lateral curve is used, invert the rules. Also make sure you know the spacing A0 for the lateral. If it is not on the heading, find a thin resistive streak and determine A0 from the log using information from Figure 2-6.

LATERAL CORRECTIONS (GENERAL)

Corrections for lateral curves are available from many sources. The apparent readings which are used to enter the charts may be one of three.

> Rop (used primarily by H. Guyod) where the value is obtained when the tool is in the center of the bed. (See Figure 2-14)
> Rat where the optimum value is used from the curve as explained in Figure 2-12 and 2-13 (used most generally)
> R_L (or R_{peak}) where the value is read at the bottom of the bed where the resistivity is a maximum or at its peak value.

All of these are used somewhere in this text. For thin beds where the 2/3 rule is used, the peak value is the easiest and most consistent.

BOREHOLE CORRECTION

Figure 2-15 is a correction for borehole effects for an 18′8″ lateral in a thick bed. In a thick bed it is

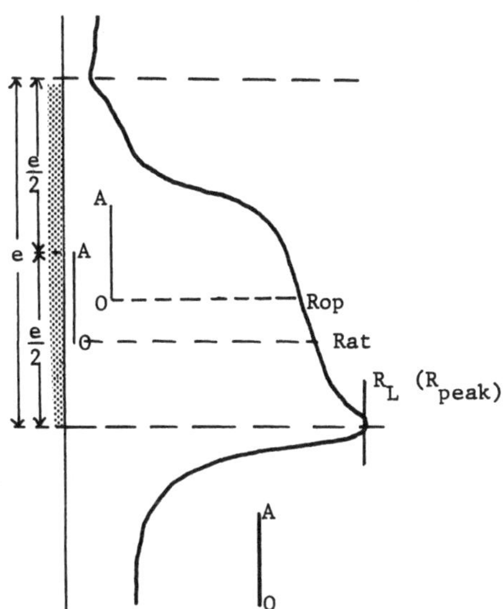

Figure 2-14. Comparison of Rop, Rat and R_L Picks from a Lateral Curve

assumed that adjacent bed influences do not exist. This Schlumberger chart is entered with Rat values. The borehole corrections, except at very high values, are all reduced when the borehole and mud values are corrected for. Invasion is not considered in this chart.

BED THICKNESS AND ADJACENT BED CORRECTION

Figure 2-16 is a chart published by Lane Wells (and is presently out of print) based on the work of Dr. L. de Witte. The chart corrects the 18′8″ lateral for adjacent bed influences (Rs) and bed thickness. Bed thicknesses down to 21 feet are correctable with this chart. The peak resistivity (R_L) is used for this chart.

The dotted curve on Figure 2-16 is for a bed with a 45 ohm m peak resistivity, an adjacent bed resistivity (Rs) of .5 ohm m and a 25 foot thickness. Enter the chart on the vertical scale with R_L and go horizontally until the adjacent bed resistivity is reached (Rs = .5). Once you reach the Rs line, go down until you reach the bed thickness line (25 feet for this case). From the bed thickness line, go horizontally until you again reach the Rs line and then go down and read the corrected value for the lateral curve (39 ohm m in this example).

It should be noted that there are really no good correction curves for 18′8″ laterals in beds less than 20 feet thick. These corrections are generally very large and should be considered as qualitative.

Most of the available correction charts are for the more modern 18′8″ laterals, although techniques such as those used to obtain Rat are applicable for any lateral spacing.

MORE COMPLEX LATERAL CURVE SHAPES

The curves presented in this section are for more complex conditions than previous examples. These are the result of runs on an analog model developed by Hubert Guyod.

Figure 2-17

This is the curve for a lateral with an A0 spacing approximately the same as the bed thickness (e). Although the bed in question is 6 times more resistive than the adjacent beds, the lateral curve shows only a modest increase in resistivity. Note the low reading just below the bed as the curve tries to produce a dead zone. Also note the reflection peak, or more correctly

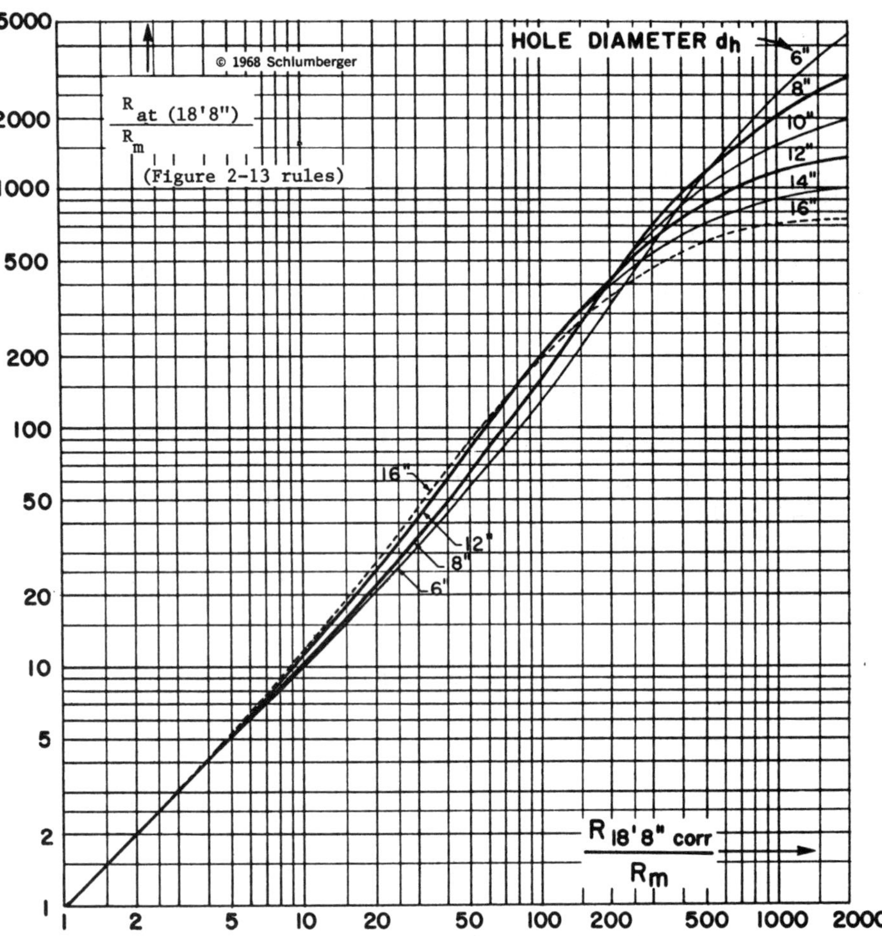

Figure 2-15. Borehole Influence Correction for 18′8″ Lateral (courtesy Schlumberger)

zone, just below. As with all thin beds, the reflection peak is about the same thickness as the bed proper. The numbers to the right of the log are the true resistivities (Rt) of the beds.

Figure 2-18

This is a lateral curve for a single resistive bed 100 times the resistivity of the adjacent beds. The bed thickness is about 60% of A0. The lateral curve shows the resistive bed, the dead zone below and the reflection zone below the dead zone. The maximum resistivity at the peak is only about 6 ohm m, although the Rt is 100 ohm m. Any correction for bed thickness would have to be about a factor of 17 times the apparent reading.

Figure 2-19

The lateral A0 spacing is about 5 times the bed thickness. The bed resistivity is very high (1000 ohm m). The curve shows the bed and the dead zone below. The reflection also shows up an A0 distance below the bed. The dead zone approaches zero resistivity as the contrast between the bed and the adjacent bed is large. This very low dead zone resistivity indicates that Rt is much larger than the peak resistivity shown.

Figure 2-20

This single resistive bed is about 90% the thickness of A0. The curve looks something like a thick bed with the slow return of the curve at the bottom of the bed. Only a small notch appears where the dead zone would start on a thin resistive bed. The peak value is about 13 ohm m where Rt is 30 ohm m.

Figure 2-21

Figure 2-21 is a single conductive bed with a thickness of about 90% of A0. Notice the trailer below the bed. Ra is very close to Rt.

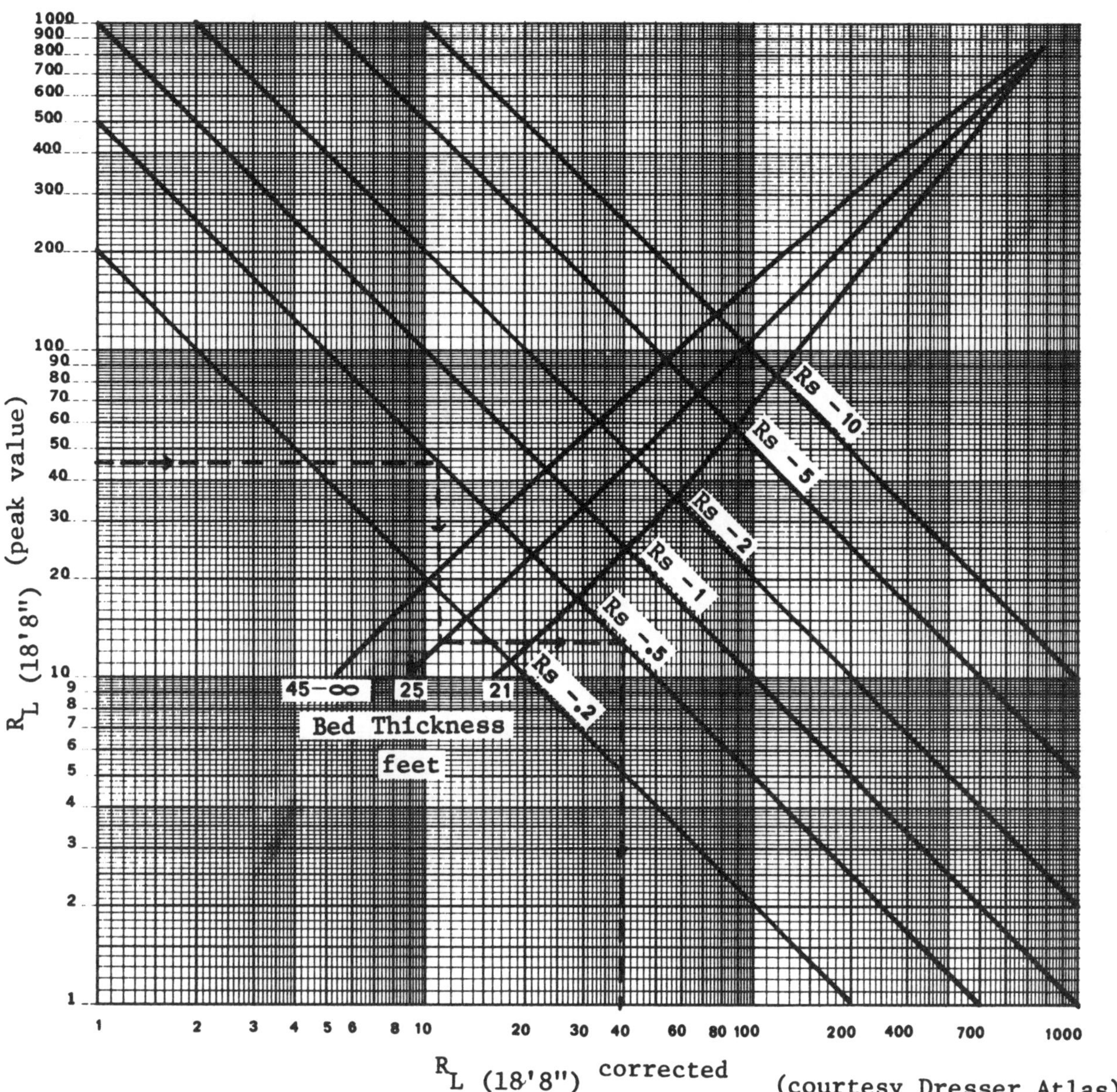

Figure 2-16. Bed Thickness and Adjacent Bed Correction for 18′8″ lateral

Figure 2-22

This curve shows a conductive bed with a thin, high resistivity streak just above it. The bed thickness and A0 are equal in this case. The trailer has a little higher resistivity as the reflection peak from the thin, high resistivity streak arrives at the bottom of the conductive bed. The Ra for the conductive bed is about Rt and only the upper 20% of the bed is not available for evaluation due to the resistivity hangover from the thin resistive bed.

Figure 2-23

This figure shows two resistive beds separated by a low resistivity bed. The bed between resistive beds a and b is about 90% of A0 in thickness. Beds a and b are both thin and will create reflection peaks. The reflection peak from a coincides with the primary peak from b and thus b has an apparent higher resistivity. The reflection peak from b shows up at c. The dead zones below a and b are not obvious.

Figure 2-24

Figure 2-24 shows a thin, infinitely resistive streak on top of a lower resistive streak with a low resistivity (one ohm m) bed above and below. The thin, upper resistive streak shows up on the lateral as a slight increase in apparent resistivity (a) with an attempt at a dead zone below (b) and the reflection peak should

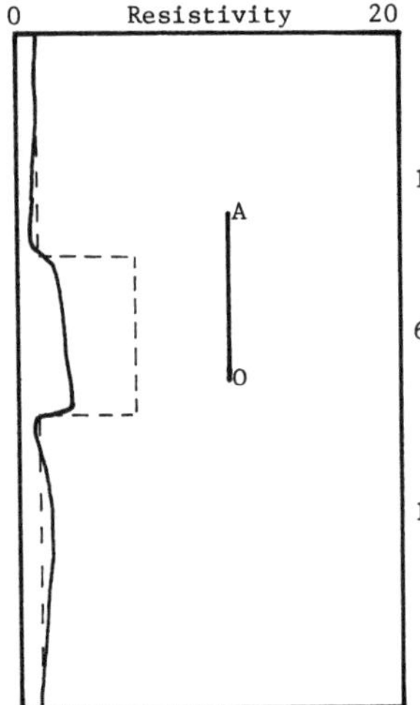

Figure 2-17. Single Resistive Bed (after H. Guyod)

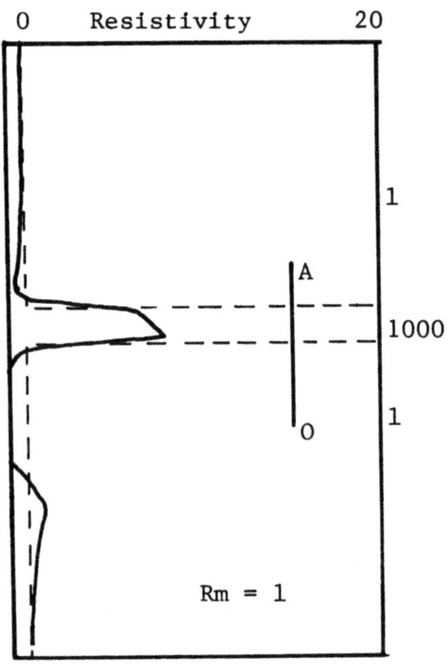

Figure 2-19. (after H. Guyod)

coincide with the bottom of the resistive section (c). The shape of the lateral curve indicates that it sees the two beds as one thick resistive bed. The hangover resistivity at d is indicative of a thick bed. This would make the total thickness 1.2 A0 which would have a peak resistivity less than Rt (about 35 ohm m). The peak is actually 60 ohm m. This means the reflection peak is superimposed on the bottom of the bed peak (c).

Figure 2-25

Figure 2-25 is a complex grouping of thin resistive beds. The group has a total thickness of A0. Primary peak a has a reflection at e, primary peak c has a reflection peak at f and b and d are probably efforts at dead zones. c has a lower apparent resistivity than a because it is in the dead zone of a. The resistive peak (apparent) at e is a combination of the reflection of a and the bottom of the total resistive thickness.

Figure 2-26

The high resistivity group of beds in Figure 2-26 is 1.6 A0 in thickness and thus will act like a thick

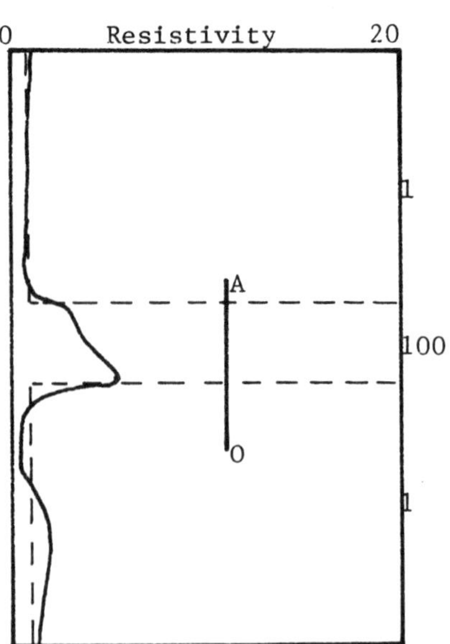

Figure 2-18. (after H. Guyod)

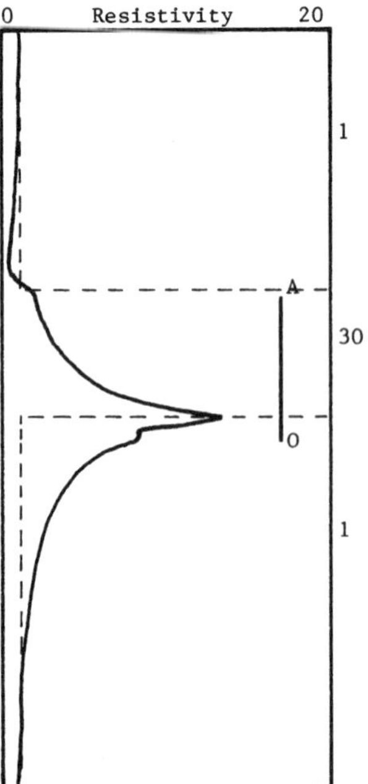

Figure 2-20. (after H. Guyod)

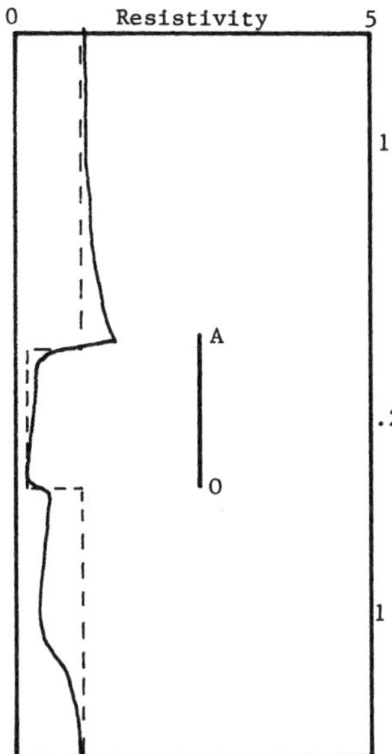

Figure 2-21. (courtesy H. Guyod)

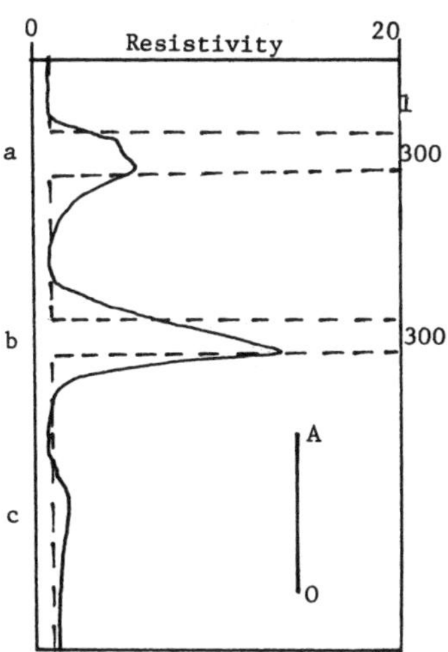

Figure 2-23. (courtesy H. Guyod)

resistive bed with a peak at the bottom. The thin, more resistive stringers in the bed will cause the problems shown. Primary apparent resistivity peak (a) is reflected at (e). Primary resistive streak (c) is reflected at (f) at the bottom of the resistive group. Primary resistive bed (f) is reflected at (g). (c) is lower than (a) as it is in the dead zone of (a). This dead zone is shown by low apparent resistivities at (b) and (d). Peak (f) is the combination of the reflection of (a), the primary effect of the resistive zone at (f) and the normal resistivity peak at the bottom of a thick resistive bed. A dead zone is evident between (f) and (g).

Figure 2-27

Figure 2-27 shows a thick (4.8 AO) bed with a thin, highly resistive stringer in the center. The upper part of the resistive bed looks normal with the notch at the top of the bed, the decay zone (a) and the buildup of resistivity. The thin, highly resistive bed (b) shows a primary apparent resistivity peak with a reflection at (d) and the dead zone at (c). The total bed shows the typical peak at the bottom of the bed (e).

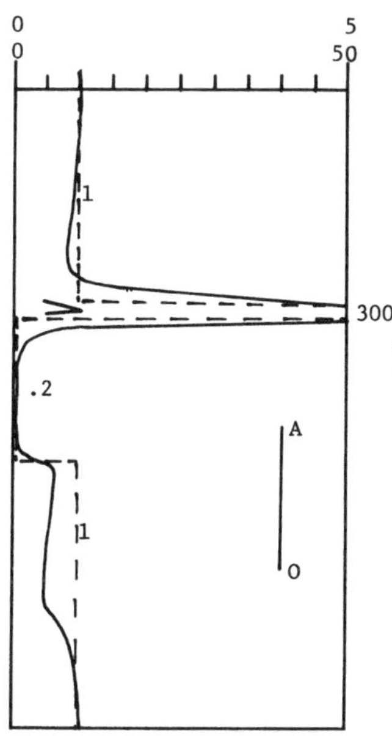

Figure 2-22. (courtesy H. Guyod)

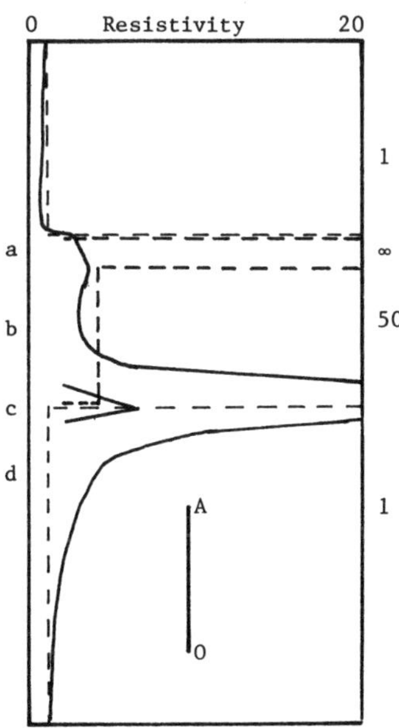

Figure 2-24. (courtesy H. Guyod)

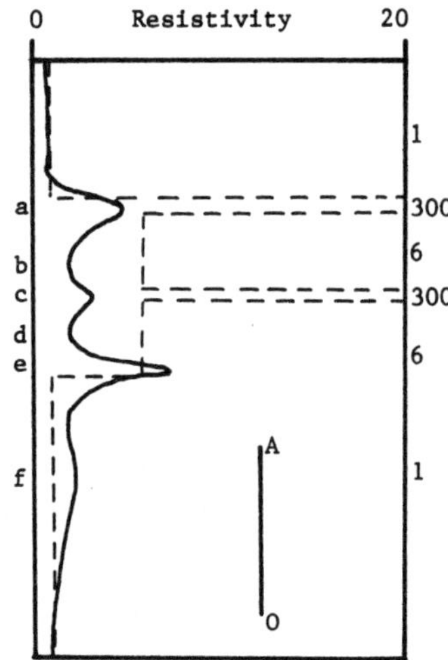

Figure 2-25. (courtesy H. Guyod)

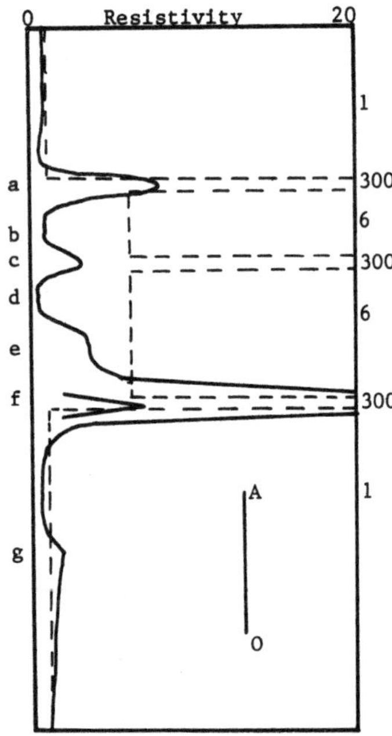

Figure 2-26. (courtesy H. Guyod)

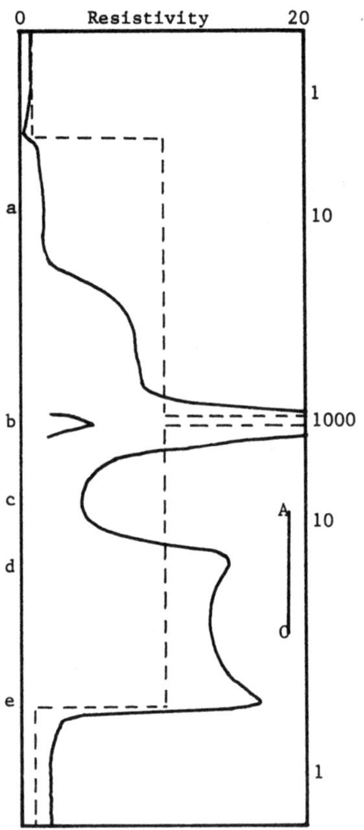

Figure 2-27. (courtesy H. Guyod)

This series of thin and complex resistive beds show the difficulty of lateral curve interpretation. In thin resistive beds even picking Rat values is often very difficult. These examples show the need to know the AO spacing and the typical curve shapes to interpret lateral curves.

More discussion of the lateral will occur in Chapters 4 and 5 where the lateral curve will be used in conjunction with the rest of the curves making up the Electrical Log.

3

Hilchie, D. W., The normal curve, Old electrical log interpretation, p. 21–27.

The Normal Curve

The normal curve was introduced by Schlumberger in 1931 to overcome the asymmetrical problems of the lateral curve. From 1932 to 1938 the normal curve became the prime resistivity curve with the lateral being used as an optional source of data.

THE NORMAL CURVE

The normal electrode array shown in Figure 3-1 is a special case of the lateral electrode array shown in Figure 2-2. The N (potential measuring) electrode is moved a very long distance away. Initially, the system was referred to as the AM∞ array. This presently is shortened to AM.

The normal measures the material between the M and N electrodes just as the lateral and the mud cup do. All the material between M and N does not have equal influences. The material close to the sonde has more influence than that far away. Figure 3-2 shows a schematic of the voltage drop versus distance from the tool. The distances are in AM distances rather than the normal feet or meters. The steeper the voltage curve, the more the influence of the material in that position. From the AM = 1 to AM = 2, the voltage drops from 100 to 50% of that seen by N. From AM = 2 to AM = 3, the voltage drops from 50 to 33% which is less than from 1 to 2. The voltage drop rate of decrease continues to reduce the further away from the tool you go. Between 6 and 7, there is little voltage drop and, thus, little influence on the resistivity measured. The depth of investigation of the normal is usually assumed to be 2AM distance, as this is the distance it takes to obtain a 50% voltage drop and represents one-half the material being measured.

Normal curves have had many spacings over the years. Modern Electrical Logs use two normals, a short normal (usually 16 inches AM) and a long normal (usually 64 inches). The short normal usually measures in what is considered the invaded zone, while the long normal reads deeper. The spacings used have been many. A more or less complete list of spacings are AM =: 8, 10, 16, 18, 38, 39, 40, 63, 64, 79 and 84 inches. Modern logs used mostly 16 and 64 inches with some locals using 10 and 40 inch spacings. Halliburton used 18 inches and the early Schlumberger log used 79 inch (2 meter) spacings. The more common spacings and combinations will be discussed as a function of time and service companies in Chapter 4.

THE POINT ELECTRODE

The point electrode was used by both Halliburton and Lane Wells for some limited time. It was run as a replacement for the short normal. The point electrode system is a special case of the normal. The current electrode and the potential (M) electrode are the same. The spacing is thus the diameter of the electrode (about $3^1/_2$ inches). The depth of investigation is still 2 × AM (as for the normal) which is now less than a foot. The advantage of the point electrode was that it had better bed resolution. The major disadvantage was that it was severely influenced by the borehole. The resistivity scales on most point electrodes is questionable.

NORMAL CURVE SHAPES

The curves for normal devices are symmetrical and thus easier to use than the lateral. Figure 3-3 shows a normal curve shape for a thick resistive bed. The dashed line is for no borehole present while the solid line shows the smoothing of the curve due to the presence of the borehole. The bed appears to be thinner on the normal curve than it is. The two shoulders at the bed boundaries disappear as the mud resistivity decreases or the borehole enlarges. In general they are not obvious. The shoulder is AM thick. The bed boundaries (both top and bottom) are found by finding the inflection point on the curve and moving $^1/_2$ AM distance away from the higher resistivity. The apparent resistivity for the curve is read in the middle of the bed.

Figure 3-4 shows the influence of the borehole on the normal curve for a resistive bed. The larger boreholes show no plateau at the bed boundaries.

As the resistive bed becomes thinner, the curve looks like the middle part of the formation is being taken away. As the bed thins, the curve does not plateau

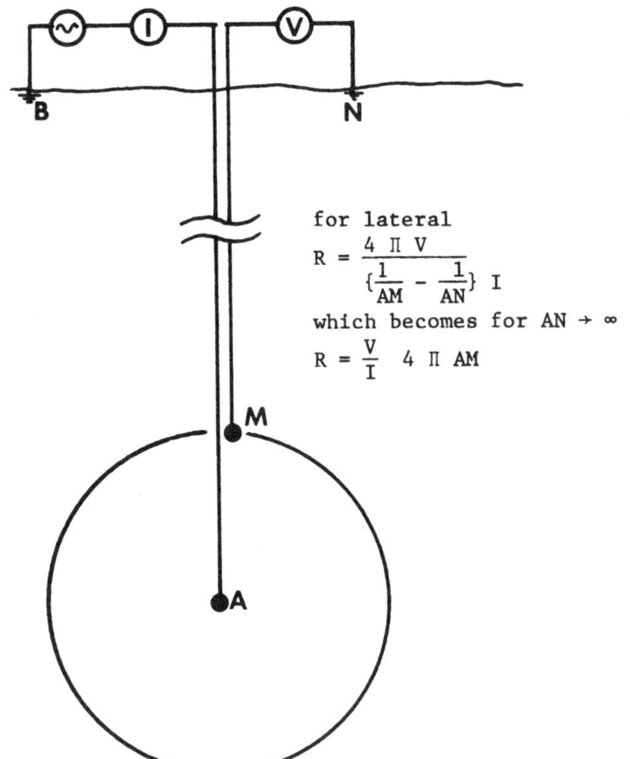

Figure 3-1. The Normal Electrode Array

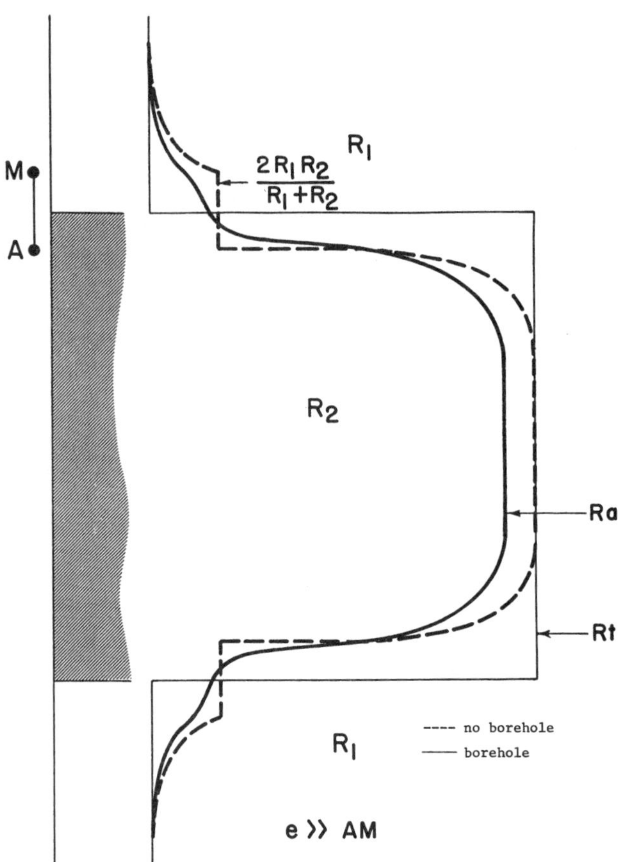

Figure 3-3. Normal Curves for a Thick Resistive Bed (after L. G. Chombart)

in the center as with the thick bed. The apparent resistivity is still read in the center of the bed which for the thin bed is the peak reistivity value. Figure 3-4 shows the normal curve shape for a resistive bed with a thickness of 2 AM. For a 16 inch normal, this would be a little over $2^1/_2$ feet. For a 64 inch normal, this would be 10-1/2 feet.

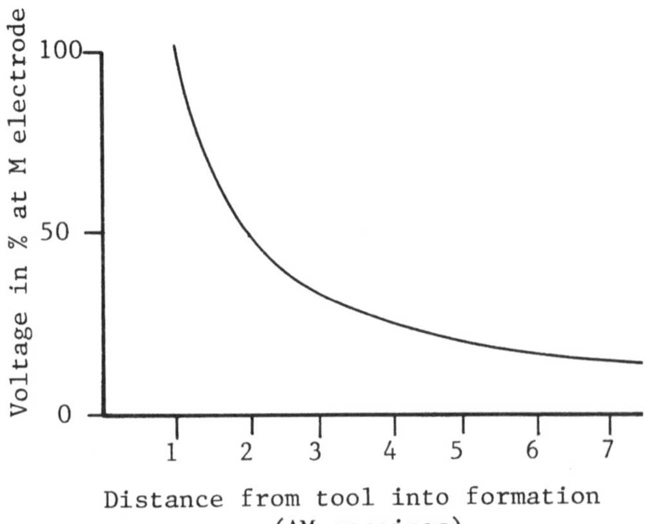

Figure 3-2. Voltage versus Distance from Normal

As the bed thickness approaches the AM distance, the peak in the center of a resistive bed essentially disappears and all that is left are the shoulders.

Resistive beds thinner than AM show up as a crater in the center of the bed with two peaks on each side. The peaks occur $^1/_2$ AM spacing outside the bed boundaries. As shown in Figure 3-6, the distance from peak-to-peak is the bed thickness plus the AM spacing. It is impossible, for beds that are resistive and thinner than AM, to determine the resistivity of the bed. All you know is that the bed is resistive and the bed thickness is the distance between the peaks minus AM.

Figure 3-7 shows the normal curve for conductive beds. Both the thin and the thick beds show in the same manner. The normal curve does a better job of measuring Rt in conductive beds than in resistive beds. The bed boundaries are found the same way as with a resistive bed. From the inflection point, move away from the higher resistivity $^1/_2$ AM distance. Conductive beds look thicker than they are due to the bed boundary effects. The apparent resistivity is

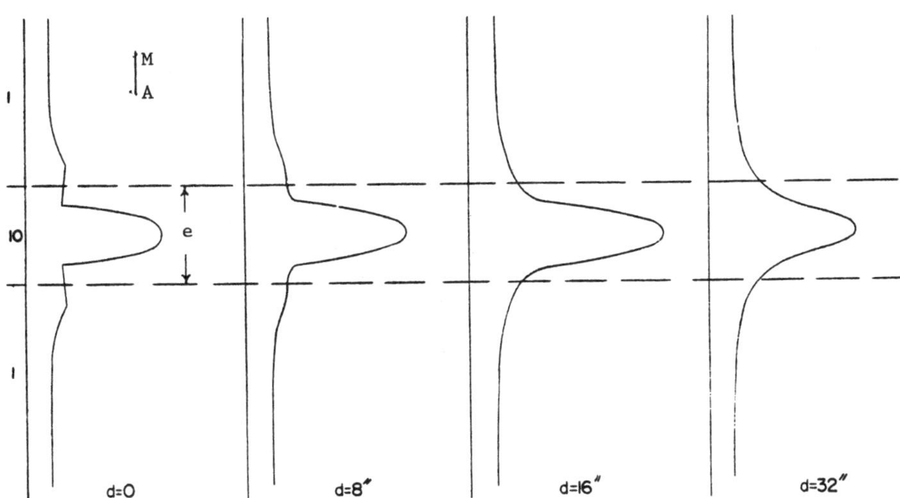

Figure 3-4. Borehole Influence on a Normal Curve (courtesy H. Guyod)

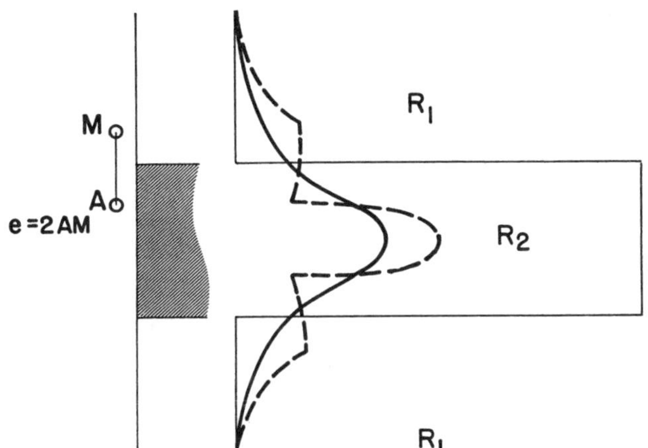

Figure 3-5. Normal Curve Shape for a Thin Resistive Bed (after L.G. Chombart)

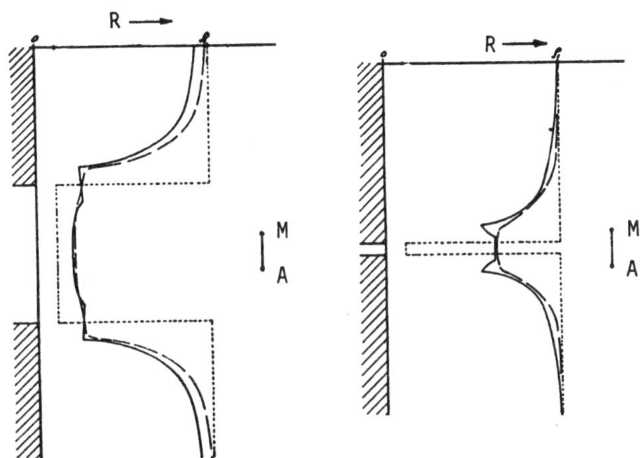

Figure 3-7. The Normal Curve Shape for Conductive Beds (courtesy Welex)

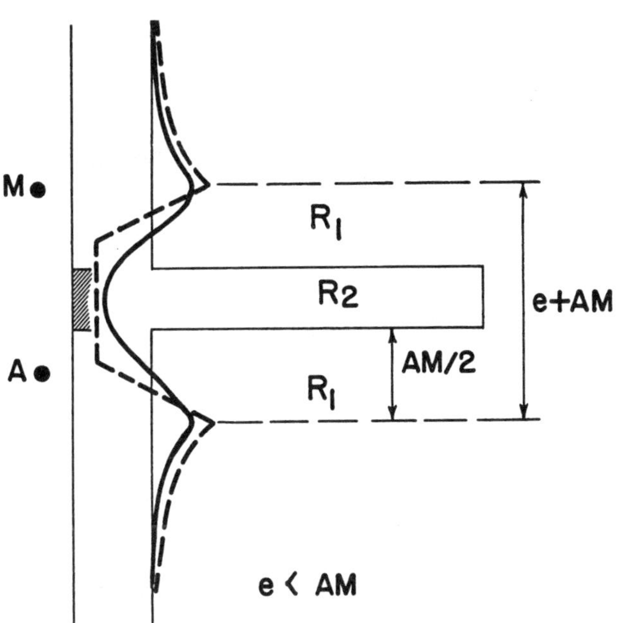

Figure 3-6. Normal Curve Shape for a Thin Resistive Bed (after L.G. Chomart)

read in the center of the bed as was the case for resistive beds.

$$Ra \simeq Rt$$

Direct readings of apparent resistivity (Ra) from the normals may not be used as approximations for Rt unless the invasion is very shallow. The short normal usually reads in the invaded zone and thus cannot be used unless there is no invasion. The long normal reads deeper than the short normal but, in my experience, does not read Rt unless invasion is very shallow. The long normal is more seriously influenced by invasion than it should be for the depth of investigation it is reported to have. A quick approximation for Rt may be obtained from the empirical relationship:

$$Rt \simeq \frac{R_{64} \times R_{64}}{R_{16}} \tag{3-1}$$

This relationship has worked well for the author.

In conductive beds the apparent resistivities are about equal to Rt.

NORMAL CORRECTIONS (GENERAL)

Corrections for the normal curves are limited to resistive beds which are at least 1.5 times the AM spacing in thickness. All normal corrections use Ra obtained in the middle of the bed where the Ra is the maximum. Bed thickness is usually obtained from the short normal (usually R_{16}) as it has the best thin bed resolution.

BOREHOLE CORRECTION

Figure 3-8 is a borehole correction for the 16 inch normal run with a Schlumberger Electrical Log. The chart may be used two ways. The first is to correct the short normal for borehole effects. To accomplish this, the apparent resistivity is divided by the mud resistivity for the particular depth and the chart is entered from the left-hand side (example shows $R_{16}/R_m = 120$). The value is carried horizontally until the proper borehole diameter (9 inches in example) is reached and then drops vertically to the scale which gives you the corrected R_{16} divided by Rm (195 for example). R_{16} (corrected) is then this latter value times Rm.

The other use is to enter the chart as before with R_{16}/R_m and go to the borehole diameter but this time the idea is to correct the value to that for an 8 inch borehole (which is common for many charts). Having reached the proper borehole diameter, go vertically to the 8 inch diameter line and then back to the vertical scale. This is R_{16} (for an 8 inch borehole) when multiplied by R_m. (Example shows R_{16}/R_m equal to 50 for a 10 inch borehole being corrected to 60 for an 8 inch borehole.)

BED THICKNESS AND ADJACENT BED CORRECTIONS

Figures 3-9 and 3-10 are charts published by Lane Wells (and presently out of print) based on the work of Dr. L. de Witte. The charts correct the 16″ and 64″ normals. These charts may be used with 10″ and 40″ normals by making approximate bed thickness adjustments shown in Table 3-1.

The correction for the 16″ normal using Figure 3-9 is performed in the same manner as the equivalent chart for the lateral curve. The example shown on Figure 3-9 (the dashed curve) starts with an Ra (R_{SN}) of 15 ohm m. The Rs is .5 ohm m. The bed thickness is 3.5 feet. From the Ra, go across to the Rs and then vertical to the bed thickness in feet. From this line

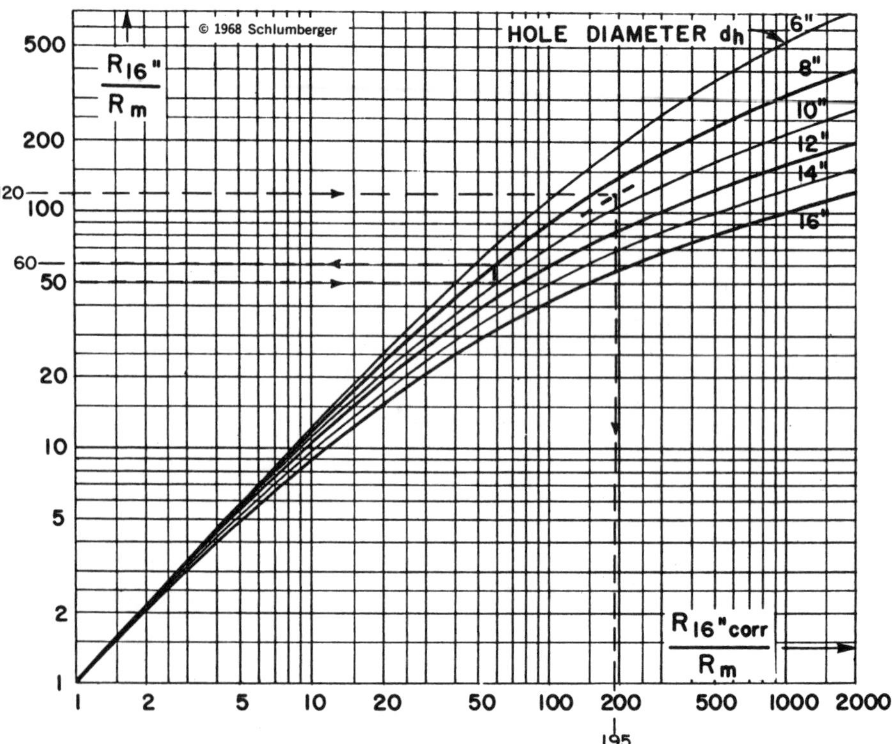

Figure 3-8. Borehole Correction Chart for Schlumberger 16 inch Normal (courtesy Schlumberger)

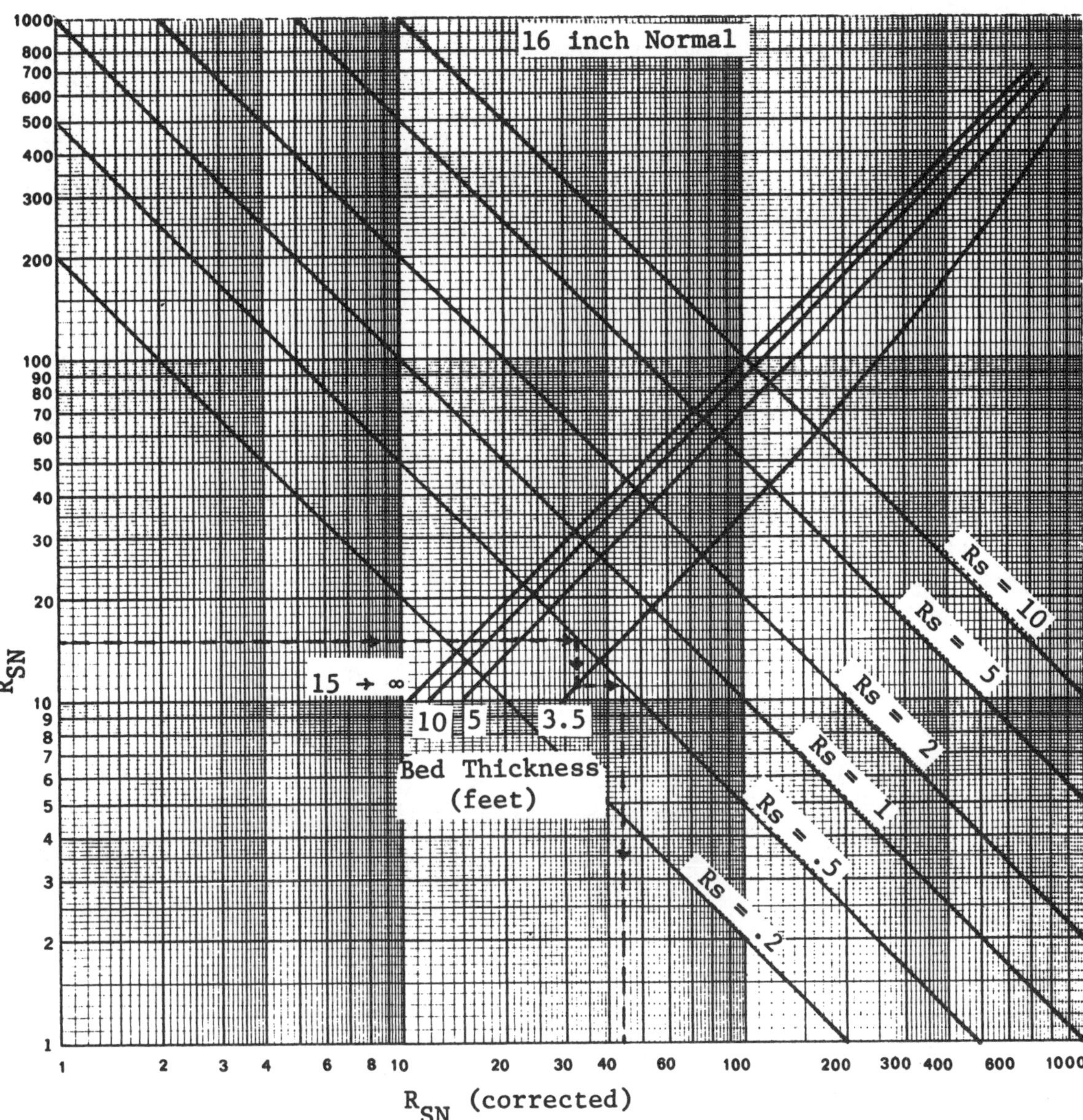

Figure 3-9. Bed Thickness and Adjacent Bed Correction for Short Normal (courtesy Dresser Atlas)

again go horizontally to Rs and then vertical to the corrected R_{SN}. Rm is not used in these charts as it is corrected for later in the departure curve section.

MORE COMPLEX NORMAL CURVE SHAPES

These more complex normal curve shapes have been taken from the analog model studies of H. Guyod. They have been chosen to present some of the difficulties in using the normals. In all cases two normals are shown, a short (relative) and a longer normal. The longer normal is usually about twice the spacing of the shorter normal. The longer normal is presented as the dashed curve. The actual resistivity is shown as numbers on the right edge of each example and shown as a curve on the simulated log. The actual resistivity line is easily recognized as it is angular and not curved. The apparent resistivities are always curved.

Figure 3-11 Normal Curves in a Layered Bed

The layered beds have resistivities of 300 and 6 ohm m alternately. The 300 beds are about the same thickness as the shorter normal. The shorter normal is thus relatively smooth with a resistivity higher than the lower 6 ohm m bed resistivities. The longer normal (dashed line) sees the 300 ohm m

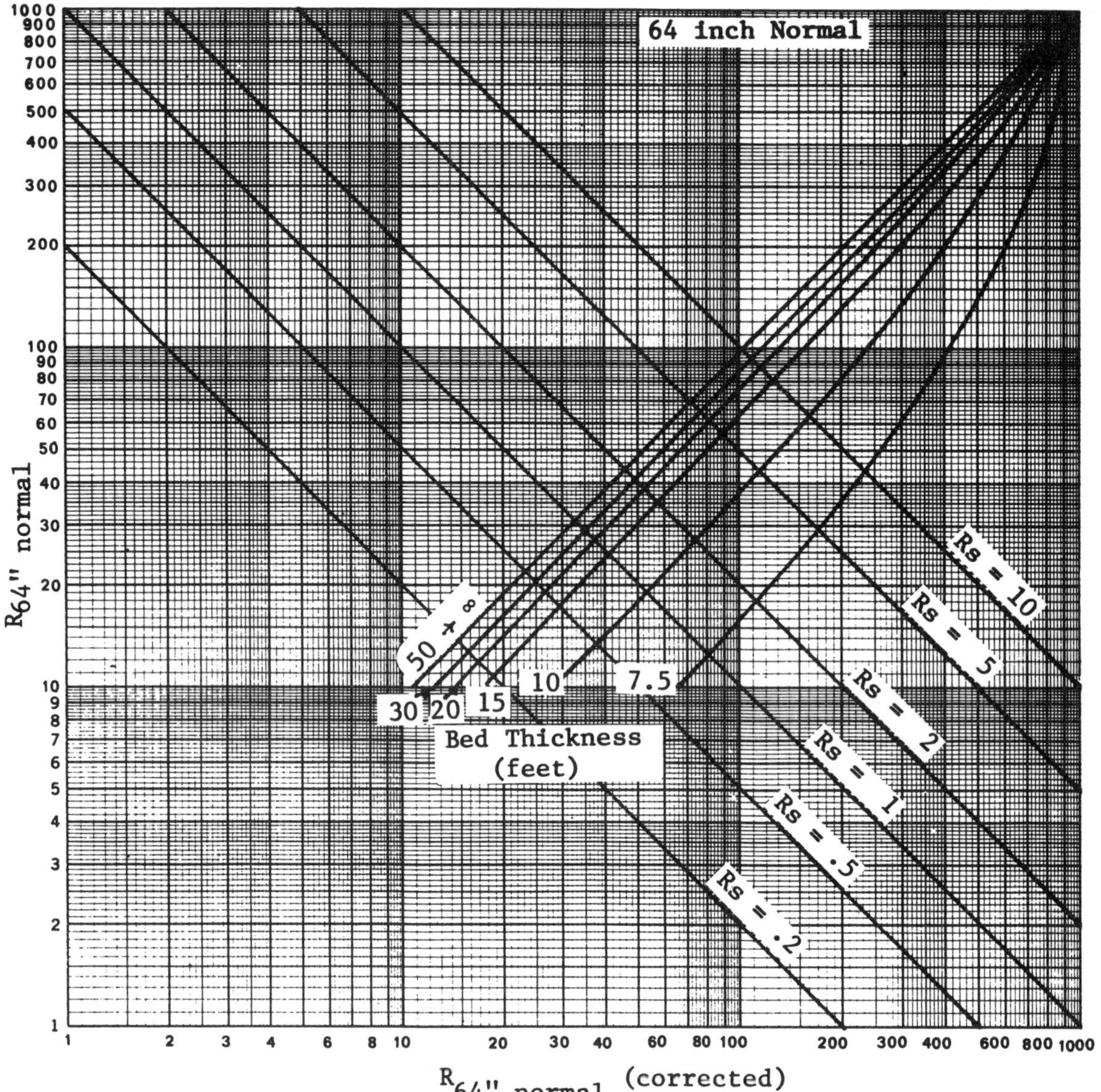

Figure 3-10. Bed Thickness and Adjacent Bed Correction for Long Normal (courtesy Dresser Atlas)

Table 3-1. Approximate Equivalents of Bed Thickness for 10″ and 40″ Normals

for 10″ Normal use 16″ Normal chart (Figure 3-9)						
actual thickness (ft.)	4.5	6.5	8.5	12	20	
use thickness (on chart)	5.	7.5	10	15	25	
for 40″ Normal use 64″ Normal chart (Figure 3-10)						
actual thickness (ft.)	5.0	7.	10	13.5	20	30
use thickness (on chart)	7.5	10	15	20	30	50

Find the actual thickness on the logs and then find the equivalent thickness for the appropriate chart from above and then use chart in normal way only with equivalent thickness.

beds as very thin resistive beds and thus has a crater opposite each of these beds. The apparent resistivity in the 6 ohm m beds is higher as these are the peaks on each side of the craters. The 300 ohm m beds at the top and bottom are almost ignored by the curves.

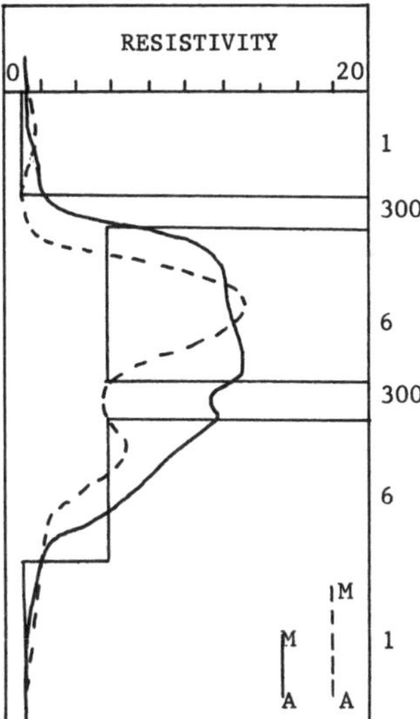

Figure 3-11. Normal Curves in a Layered Bed. (courtesy H. Guyod)

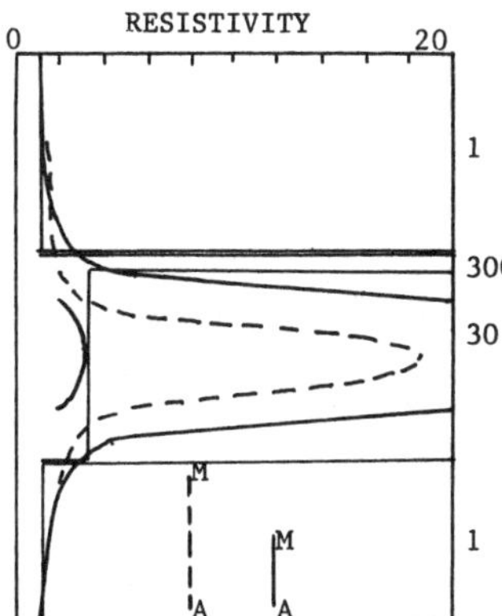

Figure 3-13. Normal Curve Shapes for a Simple Two Zone Resistive Bed.

Figure 3-12. Normal Curves in a Multiple Layered Bed

Figure 3-12 Normal Curves in a Multiple Layered Bed

The resistive bed has a 300 ohm m layer at the top and in the middle. The rest of the bed is 6 ohm m.

The shorter normal (solid line) has a longer spacing (AM) than the 300 ohm m beds and thus a crater with peaks on each side develop at these beds. The bed in the middle is obvious but the upper bed peaks are not as obvious. The resistivity of the zone between the two 300 ohm m beds is influenced by the peaks.

The longer normal (dashed) has more pronounced craters and peaks for these thin resistive beds.

It is obvious that neither of these normal curves could be used to determine the resistivity of beds next to the thin high resistivity beds.

Figure 3-13

This resistive bed has a thin very resistive stringer at the top and a thick less resistivie bed below. The longer and shorter normals both try to create a crater opposite the upper bed but don't quite make it. The longer normal does not reach an Ra close to Rt due to the bed thinness.

Figure 3-14

This example shows a thick resistive bed with a thin, much higher resistive bed in the center. The shorter normal sees the center bed as being thick enough to respond to while the longer normal sees the bed as a very thin bed and reverses,showing a crater and peaks.

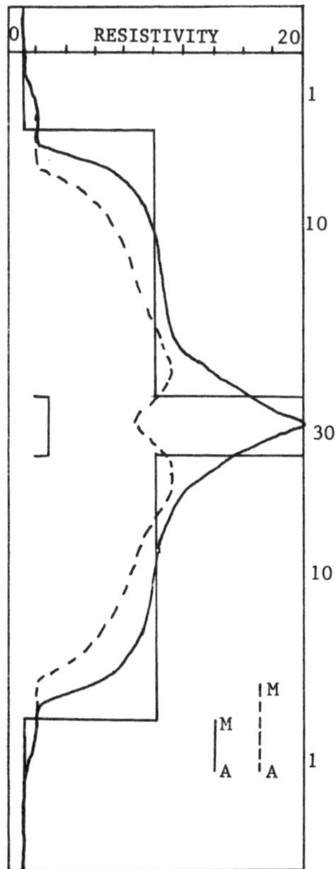

Figure 3-14. Normals Over a Thick Resistive Bed with a Tite Thin Stringer (courtesy H. Guyod)

4

Hilchie, D. W., The electrical log (ES), Old electrical log interpretation, p. 29–48.

The Electrical Log (ES)

The electrical log (often referred to as the ES from electrical survey) is a combination of serveral logging curves. These are usually the SP, a shallow investigation resistivity curve - the short normal or point electrode, and 1 to 3 deeper investigating resistivity curves - long normals and laterals. The combinations vary with time and service company.

Table 4-1 is an approximate evolution of the ES. The emphasis is on Schlumberger as they ran most of the logs. The earliest ES was, of course, just a station or point-by-point measurement of resistivity using a lateral curve (or, as was referred to in those days, an AMN). Figure 4-1 shows a copy of the first ES. This lateral is referred to as an inverted lateral. That is, it is upside down to today's version. In 1931 the SP was added to the ES. In 1932 the lateral was turned upside down and was presented the same as today's laterals. Also in 1932 the short normal was added (this was called an AM∞ curve). The spacing (A0) of the lateral changed much in this and later times. It started out as 11.5 foot A0, was changed to 7.4 foot A0 and until the 1950's varied from 7.5 to 24 foot A0 spacings. From 1932 until 1938 the SP and the short normal were the prime curves with the lateral being run as an optional curve. In 1938 the ES was standardized and more or less stayed that way until its last days.

Table 4-2 is a chronological presentation of the history of service companies that ran electrical logs. This table will assist you in making the transition of

Table 4-1. An Approximate Evolution of the Electrical Log

1927	Schlumberger first log inverted lateral - A0 11.5 feet*
1931	SP added inverted lateral A0 - 7.5′
1932	Short Normal (18″ AM) added. Lateral changed to present array. ES became primarily SP and short normal. Lateral (A0 7.5 – 24′) was optional and not often used.
1934	Long normal was added as option.
1938	Halliburton runs first ES - a single point resistivity curve
1938	Schlumberger more or less standardizes ES curves and spacings
	Most logs - Normals 16″ and 64 (some 63)″
	Lateral 18′8″
	In some areas such as Illinois and Eastern Alberta standard was
	- Normals 10″ and 40 (some 39)″
	Lateral 15′
	Until about 1947 many laterals did not conform to these standards and you see 7.5, 15, 16, and 24 foot laterals regularly. These laterals were run separately from the other curves and traced on by the engineer. About 1947 the 6 conductor armour cable appeared and all the curves could be run at one time.
	ES did not change much after 1947. Most other service companies running logs (Table 4-2) were spin-offs from Schlumberger and ran about the same set of curves. Halliburton remained the different one, running an SP, short normal (AM 18″) and two laterals. The transition from the point electrode to the short normal happened in the 1940's.

**all lateral spacings related to A0 convention; e.g., the 1931 2 meter lateral was A 2(meters)M(0.5 meters)N which becomes 2.25 meter A0.*

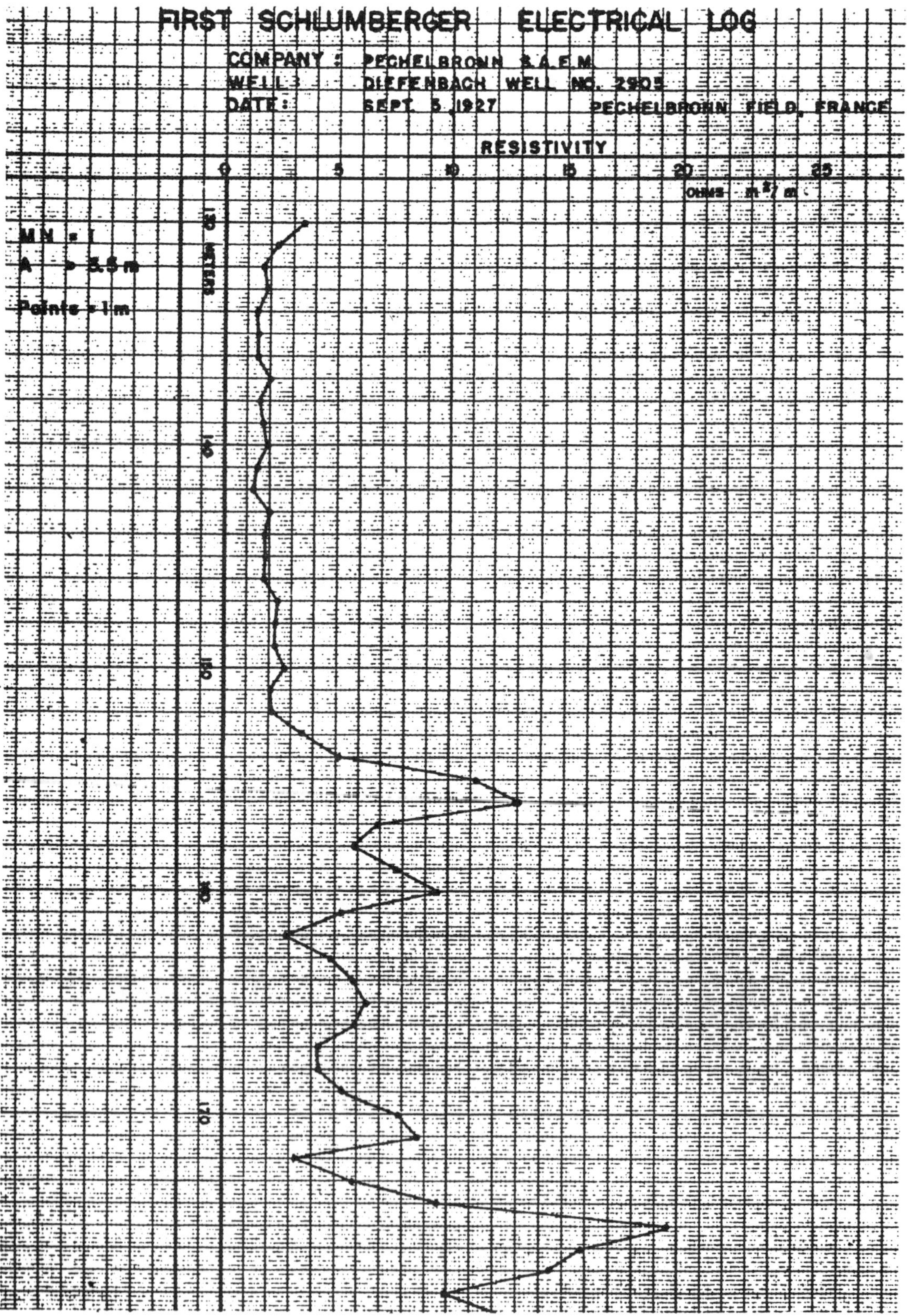

Figure 4-1. The First Electrical Log

Table 4-2. A Brief History of Some ES Logging Companies

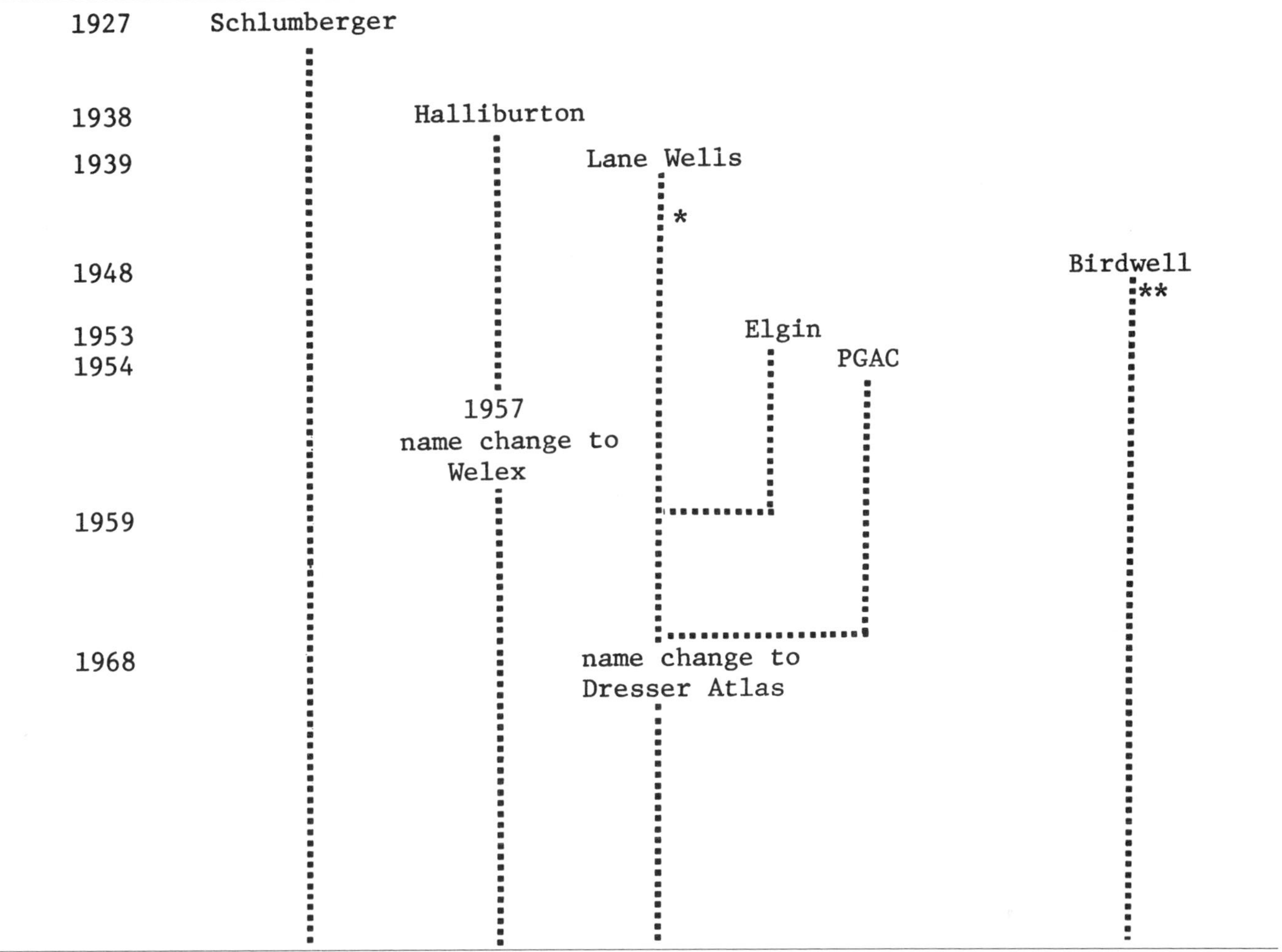

** from about 1942 to 1946 ES business inactive due to World War II Up until 1959 only activity (after 1942) was in California*
*** Started with Guard and single point electrode and in about 1951 started running ES*
Note: These dates are for ES operations only. Lane Wells and PGAC were originally perforating companies before starting ES operations.

log types and service company names. It is, of course, apparent that Schlumberger is the only consistent name in the group, although companies like Halliburton have been in the business for a long time. Most of the small companies like PGAC and Elgin were started by ex-Schlumberger personnel.

Table 4-3 is a brief summary of ES curves for various major well logging service companies for the 1940's and 1950's. All combinations available are not presented, just the most common. No effort has been made to include the curves run in California as that, in my opinion, would represent a research project all of its own.

Figure 4-2 is the standard Schlumberger sonde schematic of the 1950's. The sonde was connected to the truck by a six-conductor cable. The normals were run during one time period, followed by the lateral. These were sequenced so that the curves appeared to be continuous. The zero or depth references are shown. The differences between the short and long normal were compensated for by the camera making the log. It should be noted that the lateral uses the current electrodes straddling the 0 of the lateral rather than the M & N discussed previously. This was done to electrically balance the lines. The theorem of reciprocity makes this possible. The theorem of reciprocity says that you can interchange the potential (measurement) and current electrodes and you will get the same results. To obtain more than the three spacings that are standard for this sonde (logging tool), all you have to do is rewire parts of the system. A 32 inch normal was obtained by moving the M for the short normal from electrode #2 (from bottom of the tool) to #3.

Table 4-3. Modern (1940's and later) Service Company ES Curves

COMPANY	*SHALLOW R*	*DEEP R*	*DEEPEST R*	*COMMENT*
Schlumberger	16″ AM	64″ AM	18′8″ A0	
	10″ AM	40″ AM	15′ A0	localized
	10″ AM		18′8″ A0	w/ limestone curve
Halliburton	18″ AM Pt.	9′ A0	16′ A0	Pt. on earlier logs
PGAC	16″ AM	64″ AM 8′A0	18′ A0	Gulf Coast
	16″ AM	38″ & 64″ AM	18′ A0	Mid Continent
Elgen	16″ AM	64″ AM	18′8″ A0	Lane Wells 1960's
Lane Wells	16″ AM or Pt.	40″ AM or 14′ A0	14′ A0 or 18′ A0	1940's and in California

Nomenclature *Pt.* *point electrode*
AM *normal*
A0 *lateral*
Note: older logs and California logs may vary significantly

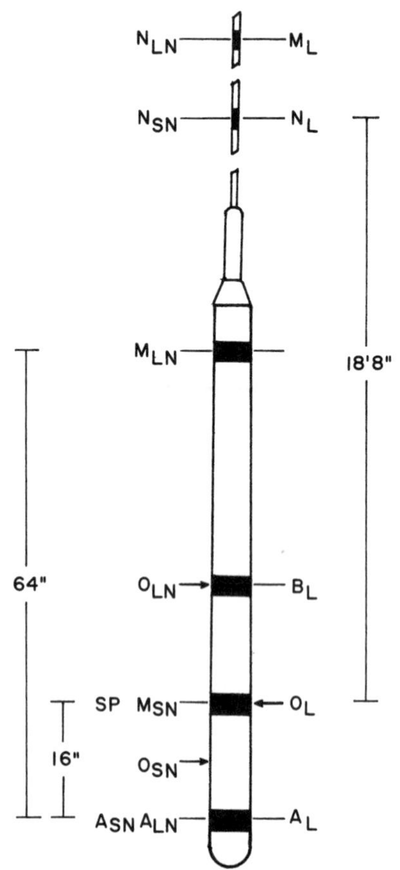

Figure 4-2. Schematic Schlumberger Sonde of 1950's

A long lateral, such as the 24 foot lateral often run, was accomplished by making a new electrode for N_L farther up the bridle (the rubber section of the cable on which the two upper electrodes are placed), connecting it to N_L. The calibrations for the curves also had to be changed. This sonde was very versatile as it contained no electronics, only electrodes and connecting wires.

Figure 4-3 is a schematic logging tool for Halliburton in the 1950's. Notice that Halliburton ran two laterals and a short normal. Halliburton referred to the curves as Z 18″ (an 18 inch normal) and 3iZ (lateral) curves. For simplification, this text uses only normal and lateral terms. The Halliburton sonde was much longer than Schlumberger's, as it contained electronics necessary to operate the FM system which transmitted the data up the single electrical conductor in the cable.

In general, the SP was always in track #1 (left of the depth column) and the short normal in track #2 (the center track). The other resistivity curves appear in tracks #2 & #3. The actual curves vary with location (see Table 4-3). The earlier logs usually did not label the curves other than curves one, two, three, etc. Generally, curve #1 was the SP which was in some logs labeled the "Porosity" curve. Curve #2 was generally the short normal. Generally, as the curve number increased, the spacing of the device increased. Most times the actual spacing and type (normal or lateral) was not indicated. To use these curves intelligently, it is necessary to determine the type of curve (point, normal or lateral) and the spacing. To do this, a good knowledge of the curve shapes is necessary. A review of Chapters 2 and 3 should

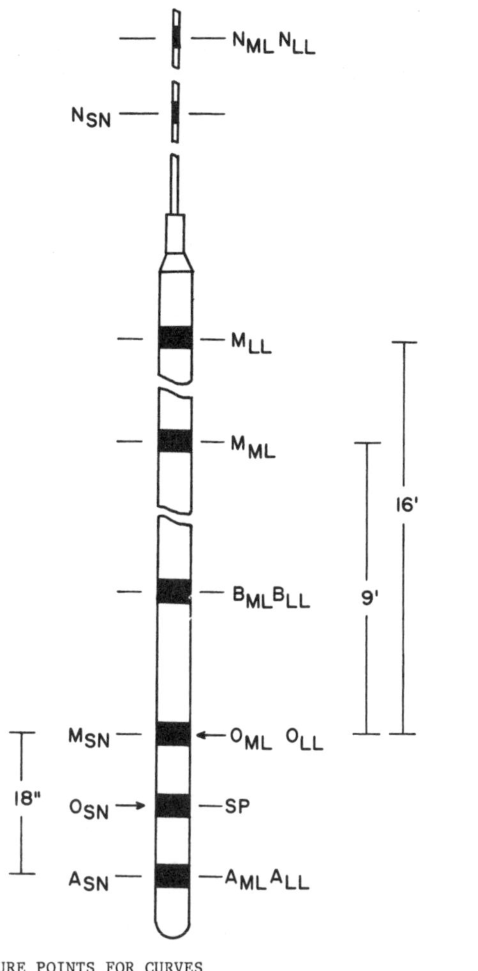

Figure 4-3. Schematic Halliburton ES Logging Tool of the 1950's

be made if these curve shapes are not readily retrievable from your mind. The lateral curve is easily recognized when there are thin resistive beds. The characteristic shape allows identification and determination of the spacing. It is usually unwise on an old ES to assume you know the curve types and spacings. The lateral, for example, was run inverted by Lane Wells in the Mid-Continental region in the early 1940's. Also, special curves were run on request by some companies.

USING THE OLD ES

In this section, I present some guidelines which help to interpret the ES. These guidelines are considered qualitative by today's logging standards but are necessary as with the old ES you seldom knew porosity, unless you have core analysis, and had to rely on what the curves would tell you and what you knew about the formation. Quantitative determination of water saturation (Sw) was difficult unless you knew porosity was constant or knew porosity from some other source (cores, drilling time, etc.). What you generally are trying to determine are: is the formation permeable, is water resistivity constant, what is porosity, and is the bed resistive enough to bear hydrocarbons? Next, I will present a review of principles directly related to these questions.

Is The Bed Permeable?

SP If you are in a fresh mud system, the SP should deflect to the left (-).
If the bed is very thin, the SP may not deflect very far
The amount of SP deflection is <u>NOT</u> an indication of permeability.

Resistivity Curves If the resistivity curves all read differently, this may be due to invasion which means the bed is permeable. See the section on Rt for more specifics.

Microlog If you have a Microlog (covered in later sections) it will tell you if the formation is permeable.

Salt Based Mud See the chapter on laterologs and microlaterologs.

Is The Rw (water resistivity) Constant?

SP If the SP peaks in the thick zones are all about the same amplitude (deflection) from the shale baseline, RW is probably constant.

Porosity?

From cores, sidewall or full.
From short normal or Microlog (see chapters on these qualitative methods)
From neutron log (see chapter on neutron logs)

Rt of Bed

There are two ways; one using departure curves which correct for bed thickness and invasion which is covered in a following chapter, or the use of values without departure curve corrections as an approximation. The latter will be used in this section as this is the most common, and if you do not have control of porosity, the extra work is not necessary. If you must guess porosity, a close Rt is good enough.

$$Ra \simeq Rt$$

1. Rt = Ra (from deep lateral) using optimum rules (Figure 2-12)
   You must know lateral A0 spacing.
2. $Rt = \frac{(R_{long\ normal})^2}{R_{short\ normal}}$ if lateral curve is unreliable.
   This is an empirical relationship that corrects long normal for for invasion effects. It has no theoretical justification other than I find it works.
3. Rt = $R_{short\ normal}$ if there is no invasion, all curves read the same, or no other curves are usable.

A good quality check on the resistivity curves is to make sure the the resistivity curves "stack" in the proper order. If there are no problem, they should stack

1. $R_{shallow} > R_{medium} > R_{deep}$ or
2. $R_{shallow} < R_{medium} < R_{deep}$ or
3. $R_{shallow} = R_{medium} = R_{deep}$

The reason is shown in Figures 4-4 and 4-5. In #1 (Figure 4-4) the invasion causes the invaded zone to have a higher resistivity than the virgin (Rt) zone. In #2 (Figure 4-5) the invasion reduces the invasion zone resistivity less than the virgin zone. Under these conditions, curves that read deeper into the formation must have a resistivity that changes from close to Ri to one approaching Rt.

Failure of the curves to stack is usually due to bed thickness effects or an uncalibrated curve.

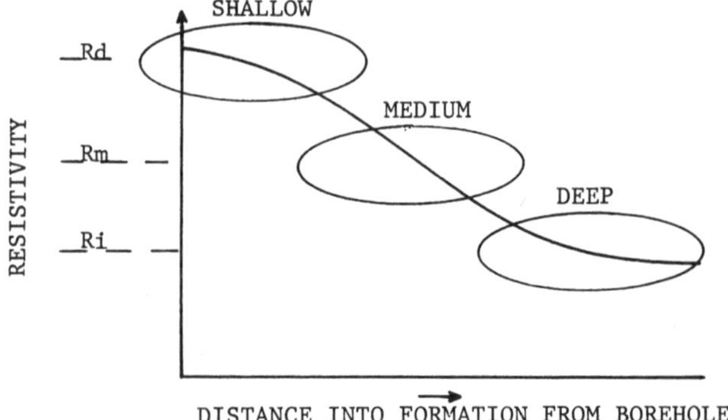

Figure 4-4. A Resistivity Profile for a Typical Fresh Mud Permeable Formation

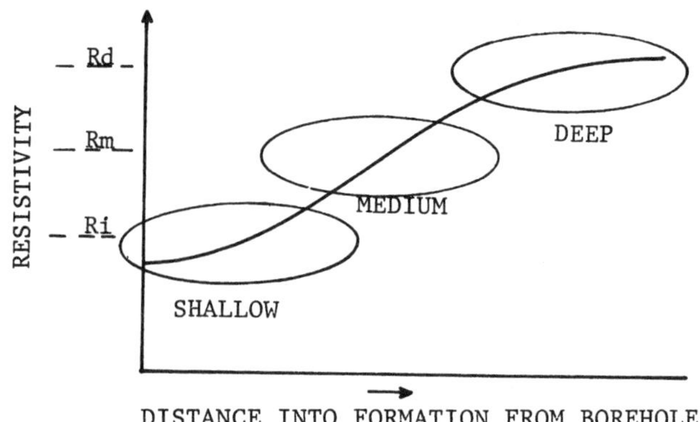

Figure 4-5. A Resistivity Profile for Some Fresh Mud and Typical Salt Mud Invaded Permeable Formation

Bed thickness checks:

If bed thickness approaches or is less than A0 or AM spacing, the reading is probably not usable.
If normal (AM) is in a bed where e (bed thickness) is less than 4 AM, the apparent resistivity may be too low.
If bed thickness appears to be a problem, correct the Ra's using appropriate charts in Chapters 2 & 3.
If the former does not work, try correcting the curves for borehole (mud) influences. This is particularly true if mud resistivity is much less than 1 ohm m.
Watch for dead zones, reflection peaks, peaks next to craters that can distort the Ra.

EXAMPLES

In this section we will look at some ES examples and show the kind of analysis that may be used. All the curves will be used together to come up with an analysis. The first is a simulated log that is meant to reinforce the simple curve shape and Ra = Rt concepts.

Figures 4-6a, 4-6b, 4-6c and 4-6d are a schematic type log for Oklahoma. I consider this a good review log. It stresses curve shapes of the normals and laterals and what to look for when qualitatively analyzing an ES. The commentary presented here is in addition to that already on the log. Cutting effect is used synonymously with decay zone. The reference will be the zone number on the log.

1. Fresh water sand 40 feet thick. The SP thickness is used in this discussion. A close analysis of the short normal would increase the thickness one foot in every case.

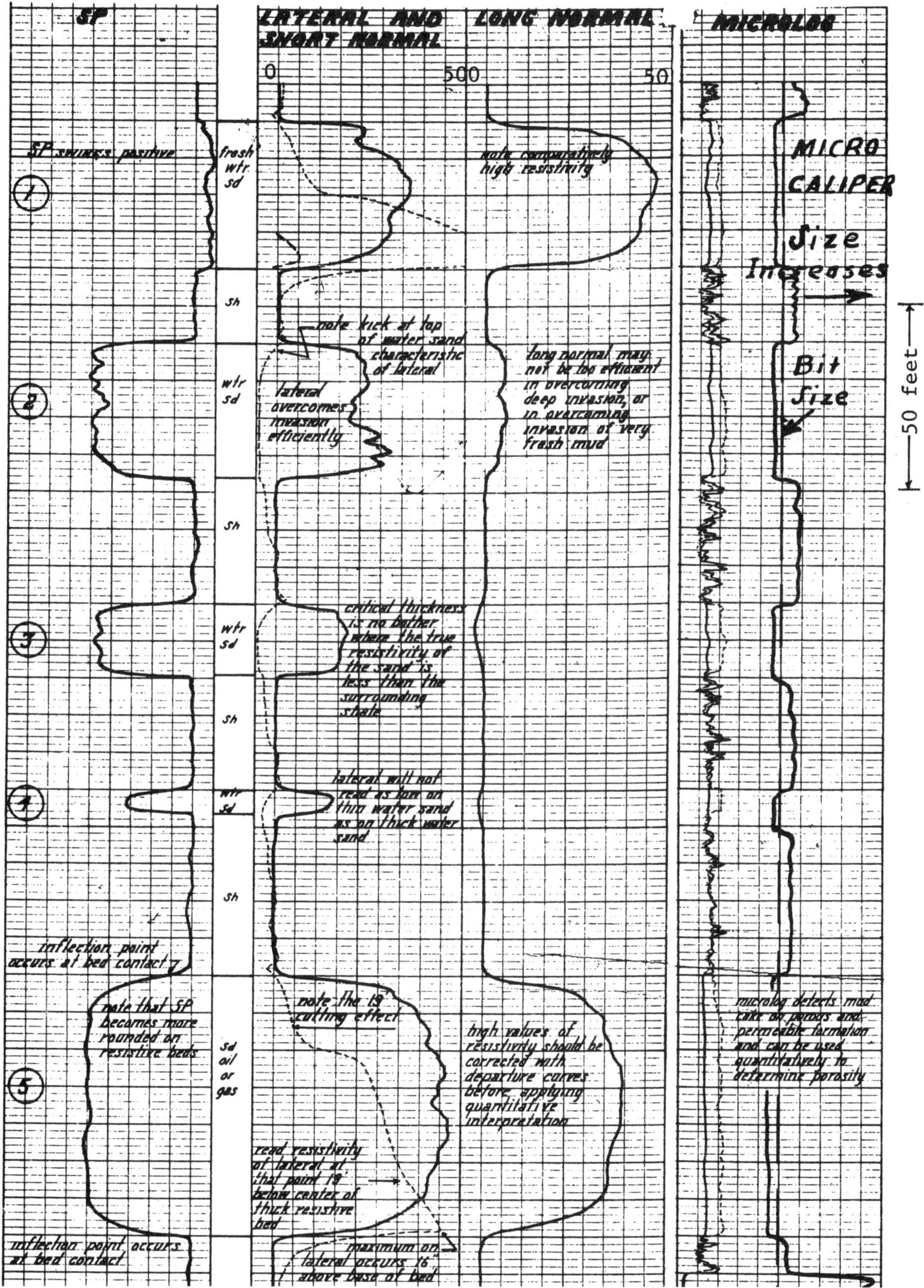

Figure 4-6 a. A Schematic ES Log With a Mid-Continent Curve Arrangement (drawn by C. K. Ruddick, 1955 with Schlumberger) AM = 16″ & 64″ AO = 19′ Rm = 1

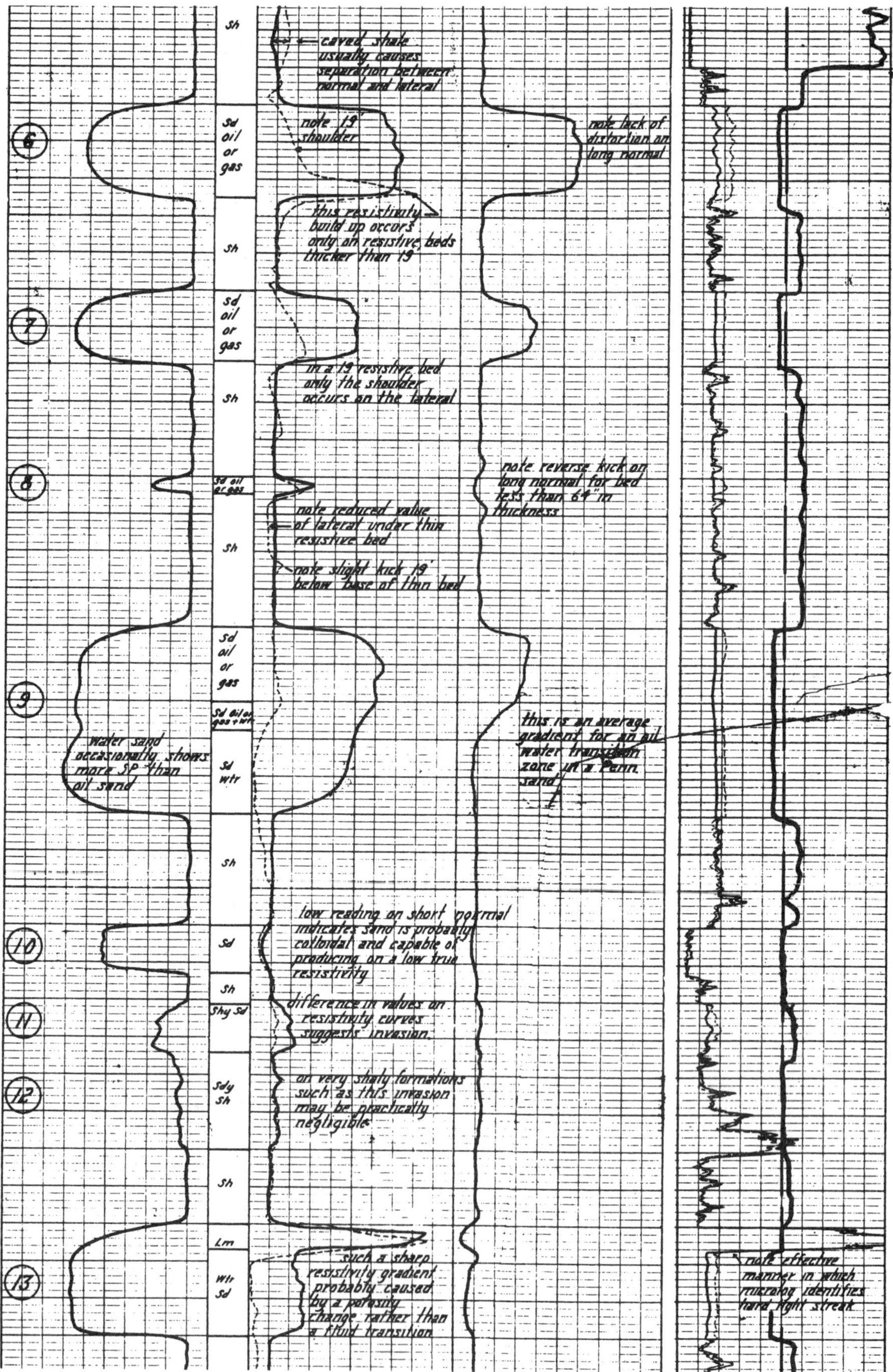

Figure 4-6 b. Schematic Oklahoma Log (continued)

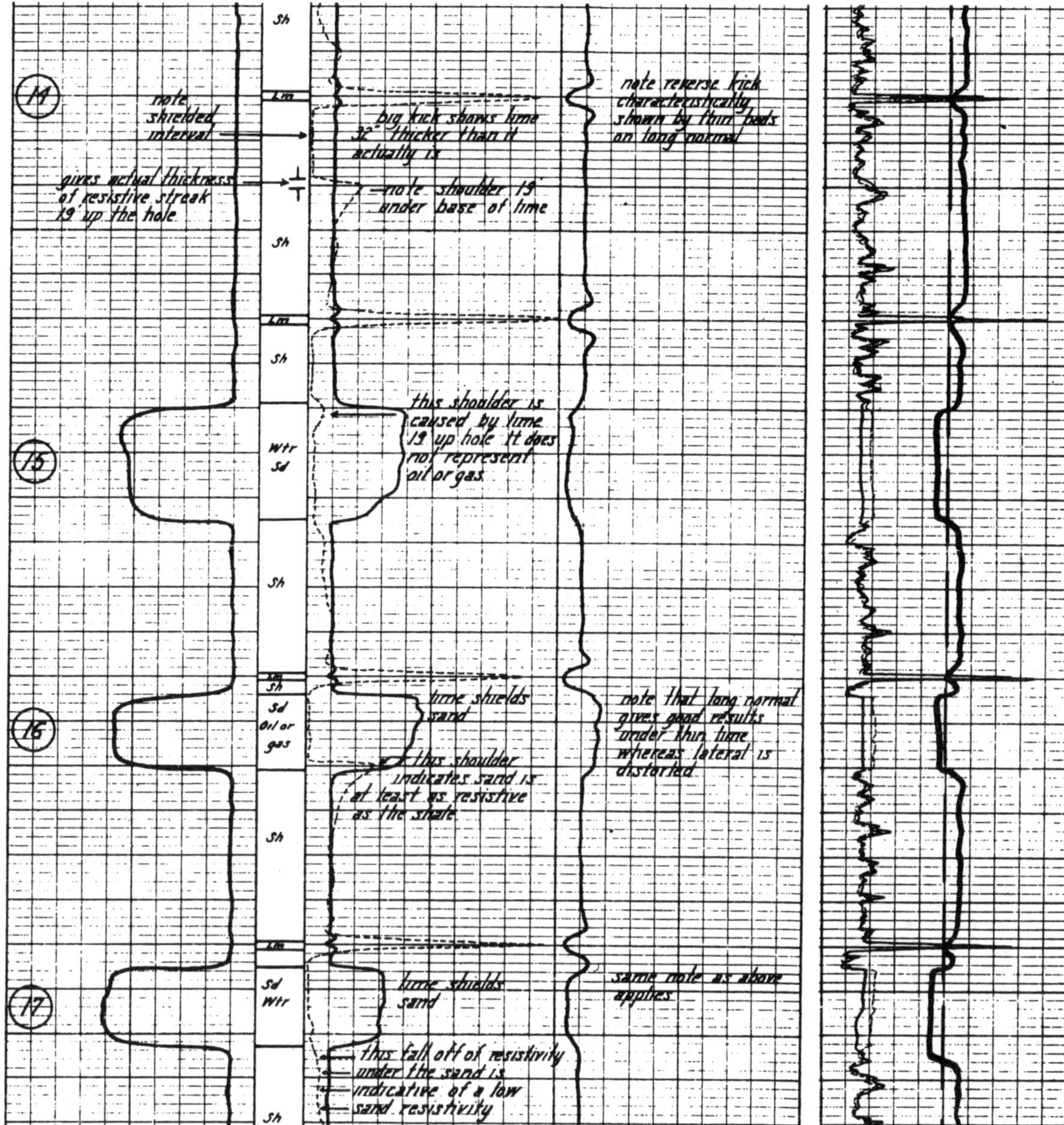

Figure 4-6 c. Schematic Oklahoma Log (continued)

$R_{SN} = 37$ $R_{LN} = 47$ R_L (use 2/3 rule) $= 2/3$ $(100 - 15) + 15 = 72$ curves stack properly

2. 36 foot bed Rmf greater than Rw
   R_{SN} 27 $R_{LN} = 10$ $R_L = 1$ curves stack properly difference is due to invasion. Note trailer for 19 feet below bed into shale.
3. 20 foot bed $R_{SN} = 21$ $R_{LN} = 2–3$ $R_L = 1$
   Just like zone #2 only thinner and with less invasion effect on LN
4. 6 foot bed. Critical thickness for a long normal in a resistive bed.
   $R_{SN} = 19$ $R_{LN} = 4$ $R_L = 2$ curve values stack properly
   Conductive bed Rt $\leqq$ 2 ohm m. SP reduced due to bed thickness. Actually SP is a little too reduced for just bed thickness.
5. 70 foot bed. Thick resistive bed.
   $R_{SN} = 45$ $R_{LN} = 39$ R_L (mid point rule) $= 37$
   Rt = 37
6. 25 foot bed. SP rounded at bed boundaries usually due to high resistivity.
   $R_{SN} = 34$ $R_{LN} = 28$ $R_L = 37$ (e = 1.3 A0 and thus use peak R) resistivities do not stack.
   Using Figure 3-10 to correct for bed thickness effect R_{LN} corrected = 35 Rt is thus around 37. Invasion from resistivity curves is not evident.

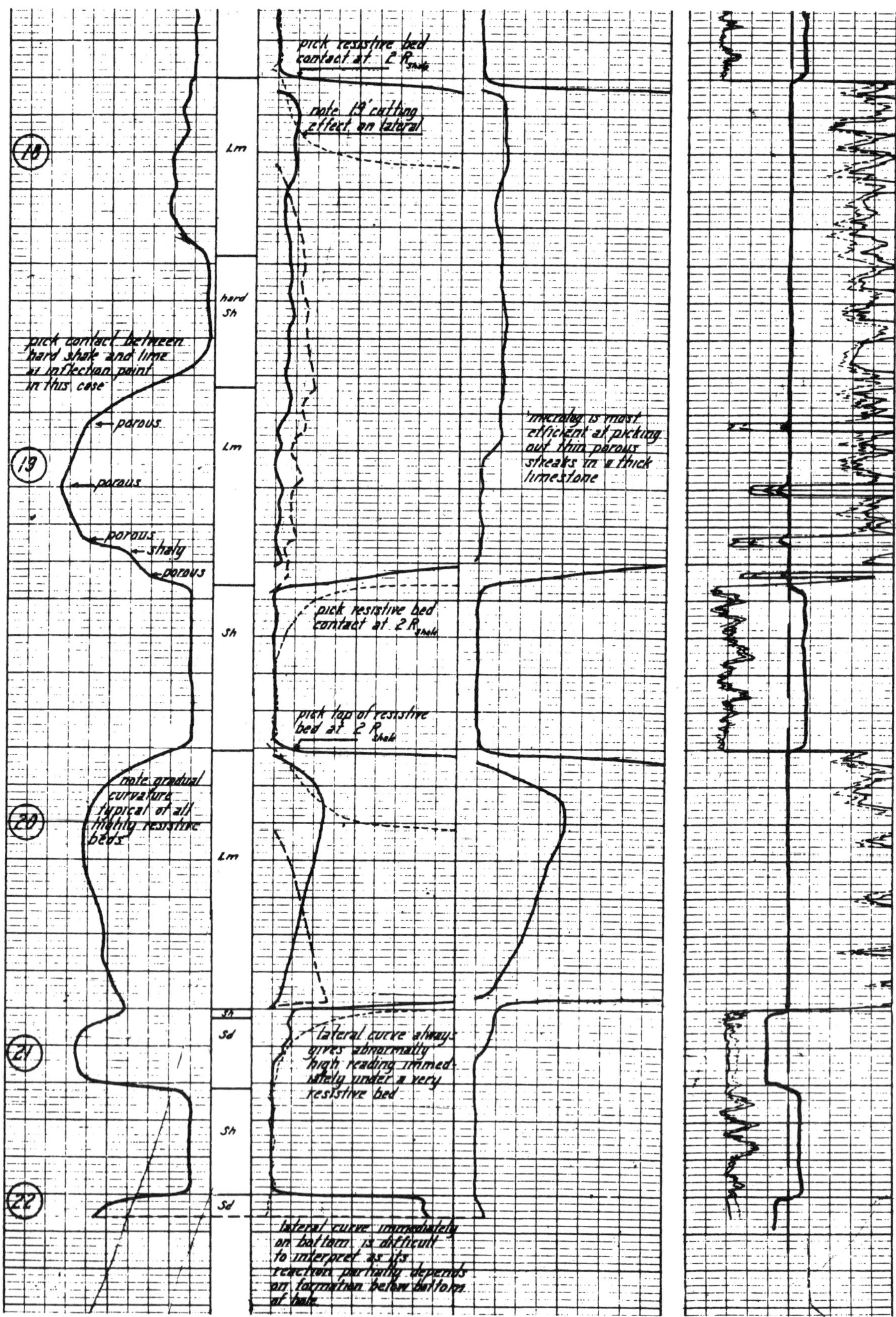

Figure 4-6 d. Schematic Oklahoma Log (continued)

7. 20 foot bed. SP is rounded so should be resistive bed.
$R_{SN} = 25$ $R_{LN} = 17$ (which corrects to 22 using Figure 3-10)
R_L should not be used as A0 is about equal to bed thickness and the bed is a resistive bed.

$$R_t \frac{R_{LN} \times R_{LN}}{R_{SN}} = 22 \times 22/25 = 19 \text{ ohm m.}$$

8. 4 foot bed. $R_{SN} = 15$ R_{LN} reversed - crater due to resistive bed thinner than AM spacing. Lateral shows bed, dead zone below and 19 feet below bed is reflection peak. SP reduced due to thin resistive bed.

$$\text{guess at Rt} = \frac{\text{Rmax}}{\text{Rmin}} \times \text{Rs} = \frac{12}{3} \times 5 = 20$$

9. 50 foot bed.
upper 20 foot section $R_{SN} = 30$ $R_{LN} = 17$ $R_L = 5$ SP suppressed over upper section, looks like hydrocarbons when commpared to lower 20
lower 20 foot section $R_{SN} = 24$ $R_{LN} = 4$ RL = 1 total section looks like 20 feet of water at bottom, 10 feet of transition zone and 20 feet of hydrocarbons at the top.
10. This is a unique situation that was noticed in Oklahoma and up into the Rocky Mountains which has been described as colloidal. Apparently, something like illite is in pores. The SP is often larger than normal and the resistivity is lower than normal. Sometimes there is no apparent invasion. Porosity can, on a density log, appear to be over 20% and the formation have no permeability. These very low resistivity zones can produce hydrocarbons and no water.
11. Permeable, Rt about 6 ohm m. Shaly.
12. Probably tite and nonproductive.
13. Tite lime stringer on top of permeable bed. Note SP and resistivity. LN is reversed and short normal shows bed as resistive. Lateral sees stringer as thin resistive bed. Lateral has dead zone for 19 feet below stringer and then reflection peak.
Permeable zone 24 feet thick. $R_{SN} = 13$ $R_{LN} = 3$ Upper 20 feet of permeable stringer is in dead zone of lateral and thus not usable. R_L at bottom of the bed 1 ohm m. Rt around 1 ohm m. Note trailer below permeable bed indicating Rt less than Rs.
14. Typical thin tite stringer. Could be lime or coal stringer. Note dead zone and reflection peak.
15. 25 foot zone. Note reflection peak on lateral at top of bed due to bed 19 feet above. $R_{SN} = 19$ $R_{LN} = 2$ $R_L = 2$
16. 16 foot zone. Lateral not usable as dead zone covers complete bed.

$$R_{SN} = 22 \quad R_{LN} = 8 \quad Rt = \frac{8 \times 8}{22} = 3$$

17. Same as #16 only $R_{SN} = 16$ $R_{LN} = 3$ Lateral again not usable.
Rt = 1 or less.
Reflection peak smaller due to low bed resistivity. Note trailer on lateral. Rt less than Rs.
18. Tite thick zone.
19. SP shows some permeable streaks (marked as porous). If permeable streaks were conductive they would show up so zones are most probably around the same resistivity as the tite zones. See Figure 4-7a below

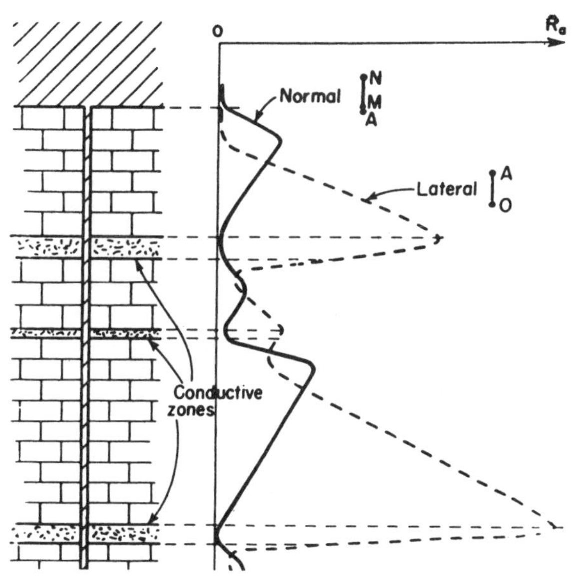

Figure 4-7 a. courtesy of Schlumberger

20. 70 foot high resistivity bed. Curve on lateral looks normal. Rt should be at least 1500 ohm m. Normals peak near top and then apparent resistivity reduces. In this case, the normal changes from an apparent two electrode device to a three electrode device due to the close proximity of N. For a modern Schlumberger short normal, this peak should be 20 feet from the top of the bed (AN = 20 feet) and for the long normal, it should be 70 feet from the top of the bed (AN = 70 feet). Curves for zone #20 may be good for old 4 conductor cable (rag line). See Figure 4-7a for another

example of normals turning into 3 electrode curves.

The maximum resistivity at peak is a function of borehole diameter and mud resistivity. The maximum a short normal can record is:

$$Rmax = \frac{8\ Rm\ AM\ AN}{d_h^2 - d_s^2}$$

for: Rm = 1 AM = 1.33 (16 inches)
AN = 20 feet
d_h = .67 feet (8 inch borehole)
d_s = .29 feet (3.5 inch sonde)

Rmax = 585 ohm m

for a 9 inch hole Rmax = 445 ohm m and for a 12¼ inch hole Rmax = 220 ohm m for an Rm = 1 ohm m.

If the short normal reaches this maximum level, it means that the bed is essentially infinitely resistive.

21. Lateral curve is hanging over from zone in #20. There is no lateral trailer below bed #21 and thus Rt is equal to or greater than Rs of 5. R_{SN} = 9 at top and reduces to 5. R_{LN} has same values. These are thus good estimates of Rt.
22. In these older days it was common practice to drill a few feet into the target zone and log. Care must be taken when evaluating the curves under these conditions. If you go to ideal curve shapes, they will help. The schematics below will diagnose #22.

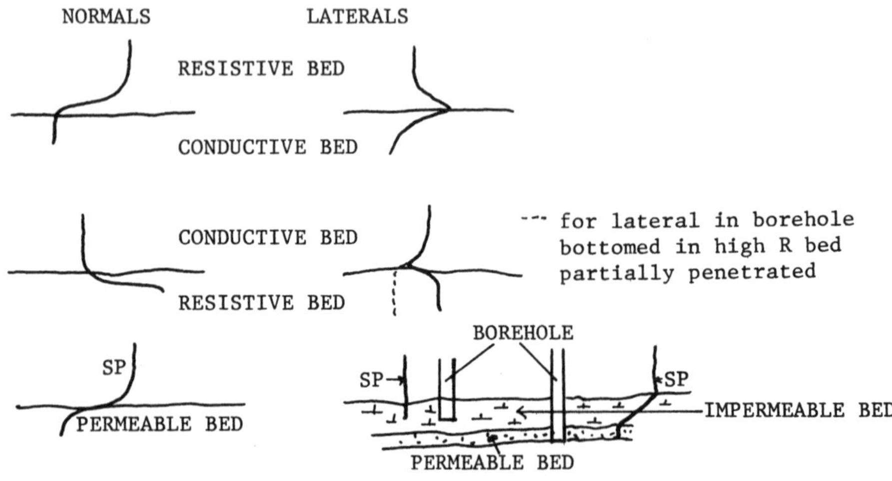

Figure 4-7b is a redraw of Figure 4-6, zones 6 through 13. From a comparison of both arrays, you will see that the Gulf Coast curve arrangement does not allow as easy an interpretation of the lateral curve but makes both normals easier to use.

EXAMPLES— REAL

Figure 4-8 is an ES from southern Venezuela. 10, 16, 38 and 64 inch normals were run and have been overlaid for comparison. Also run were a 19 and 32 foot lateral. The interval between 9550 and 9600 is an oil reservoir and the water is low salinity. This latter accounts for the very high resistivities. The shorter spaced normals have better than bed resolution but do no see as far from the borehole. Notice that the longer normals record higher resistivities as they are less influenced by the invaded zone. For example, at about 9590 the 10″ normal reads 230 ohm m, the 16″ normal reads 250, the 38″ 900 ohm and the 64″ reads 1100 ohm m. The laterals read 2400 ohm m for the 19 foot and 3600 ohm m (using the 2/3 rule) for the 32 foot. It should be noted that using the 2/3 rule can be quite inaccurate. You should notice that the curves stack properly and that the bed is not completely homogeneous (in terms of resistivity). This is a problem as with the lateral curves you obtain only one resistivity for the 50 foot bed. With the normals, more than one value can be obtained. Using the 16″ and 64″ normals and the approximate Rt relationship (equation 3-1), Rt is 4800 ohm m.

Figure 4-9 is a 1936 Schlumberger hand recorder log with just the SP and short normal. About all you can tell is that the Wilcox and Oswego are resistive beds. It appears that the backup resistivity scale is 0–250 ohm m. Many of the older logs use 5:1 scale ratios for the backup rather than 10:1 which the modern logs use.

Figure 4-10 is a Schlumberger ES probably from the 1940's. The curves are the SP to the left of the depth column (as always except sometimes in California), the short normal (as always the solid line) to the right of the depth column and a lateral. Watch the depth scales

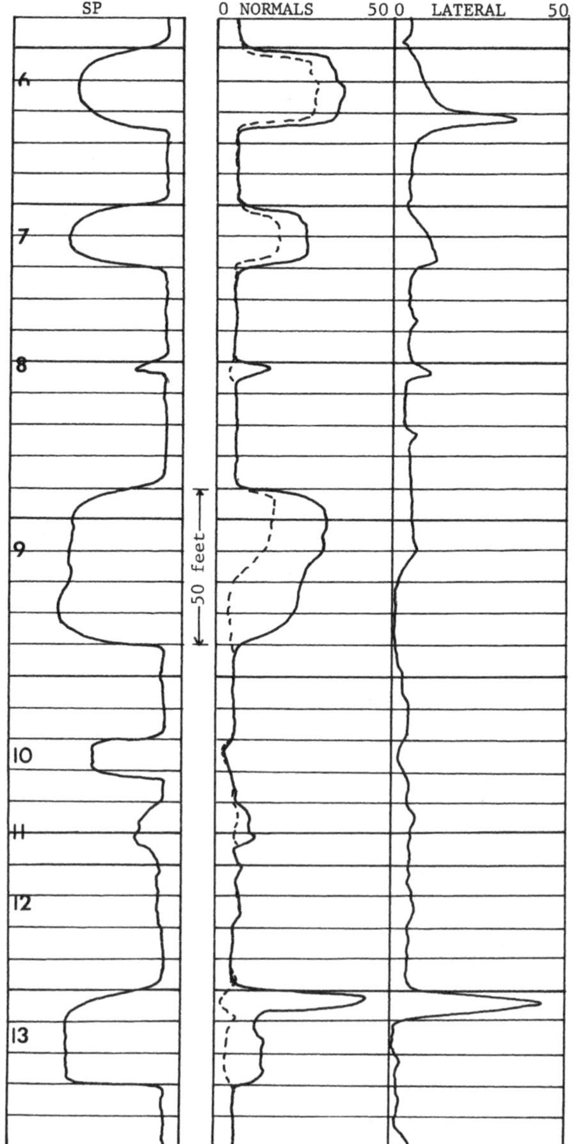

Figure 4-7 b. Schematic Electrical Log on Gulf Coast Format (zones 6–13 are the same as Figure 4-6 only on Gulf Coast format)

on these older logs. Each horizontal line is 5 feet. I find it difficult to determine the spacing of this lateral curve from this log. A survey of ES's in this time span in Oklahoma indicates that the lateral appeared to have a 15 foot A0 spacing. The upper part of the Hunton appears to be tite as the SP is severely reduced, probably due to the two shale stringers. The 30 feet above the bottom of the hole have a good SP development. From the lateral, the bottom of the resistive bed is 4025 feet which is picked as the oil/water contact. The borehole does not sufficiently penetrate the water zone to obtain an Ro for the water zone. Notice that the SP is a little better developed in the water zone.

Figure 4-11 is a Schlumberger ES of the same vintage as 4-10. The Layton sand appears to be water bearing, although caution must be used in the upper part of the sand. The high resistive, thin stringer at 2805–10 causes the lateral to be dead for about 15 feet. The reflection peak at 2823 is very small which would make you believe the zone at that depth is low resistivity. Notice the trailer on the lateral 15 feet below the permeable Layton. The short normal shows essentially no increase in resistivity at the top of the bed and thus I assume it is all water bearing.

Figure 4-12 is a modern Schlumberger ES showing an oil water contact. This example is unusual in that the short normal and long normal show no general increase in resistivity as the curves move from the water zone to the oil zone. Usually an increase is noted due to the residual oil in the invaded zone opposite the reservoir. The lateral shows a general increase from about 2.5 ohm m to 15–20 ohm m at the top. The formation is probably deeply invaded and the invasion has pushed the hydrocarbons out of the field of vision of the normals. The SP shows a definite reduction due to (probably) hydrocarbon suppression. The constant short normal is a good indication of constant porosity.

Figure 4-13 is an example of a modern ES water bearing zone. The resistivity curves all increase towards the top of the permeable zone but this is due to porosity reduction. The SP is the only key to no hydrocarbons as there is no hydrocarbon suppression.

Figure 4-14 shows a water zone at 3630–40 (each horizontal line is 2 feet) and an oil zone at 3571 – 82. The 18′8″ lateral (dashed curve in track #2) is dead throughout the oil zone due to the high resistivity thin bed above the pay zone. The short normal reads 35 ohm m and the long normal reads 44 ohm m. Using the two normal systems, Rt becomes 55 ohm m. The water zone at 3630 – 40 has the same problem with the lateral. The permeable zone is in the dead area of the lateral curve and so it is not usable. The short normal reads about 10 ohm m and the long normal 8 ohm m. This makes Rt about 6.5 ohm m. Notice the SP's. The oil zone SP is more rounded than the water zone.

Figure 4-15 is a modern PGAC ES with four resistivity curves. The zone at 7776 – 88 (each line represents 2 feet) is resistive and has an SP development. The short and long normal read almost identical resistivities. The laterals are quite different. The short lateral shows a large resistivity anomaly, while the long lateral does not. The long lateral sees the bed as

Figure 4-8. A Composite ES from Venezeula with Multiple Normals & Laterals

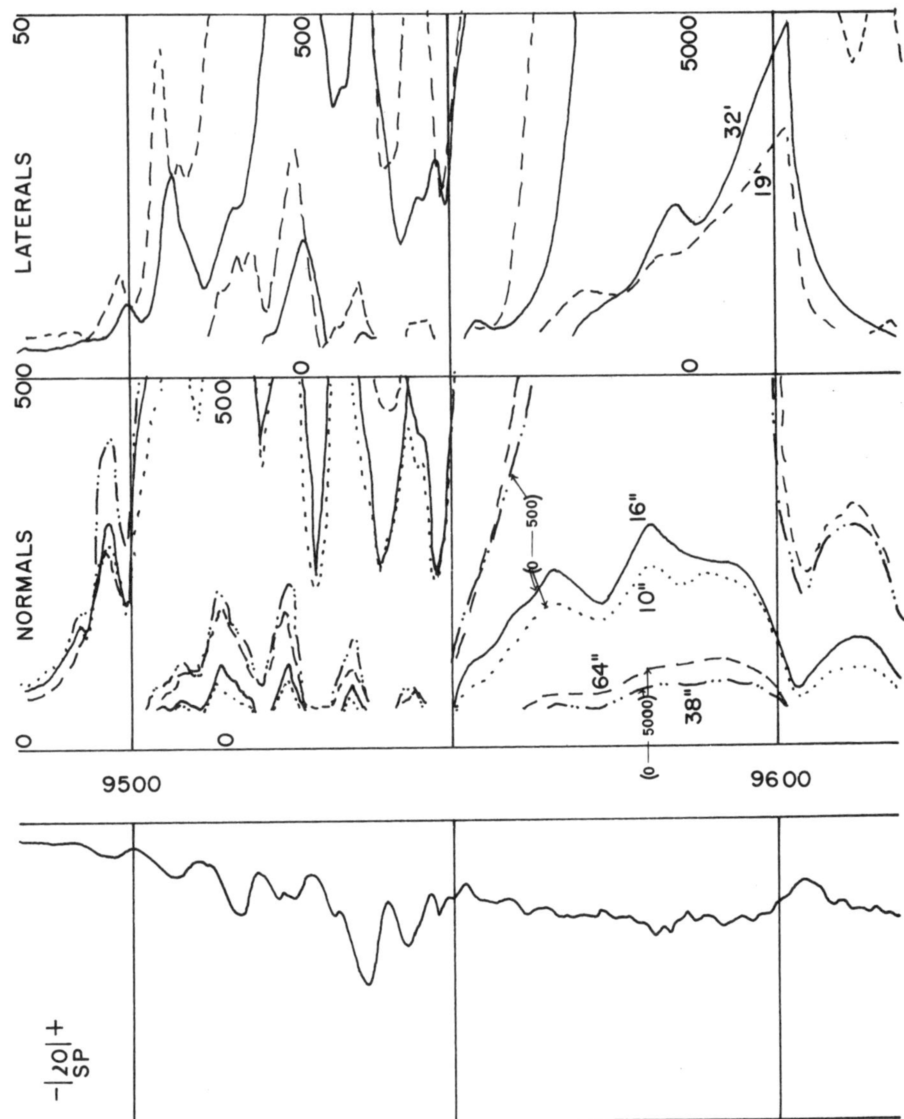

very thin and responds with a resistivity show opposite the bed, a dead zone below the bed and a reflection peak 18 feet lower. Notice that the lateral curves appear to be displaced downward in the bed due to the asymmetrical shape of the laterals. The lower resistive bed at 7818 is very thin for all curves. The short normal starts to see it, the long normal reverses over the internal making the bed thickness less than 5 feet. The two laterals see the resistive bed and have dead zones and reflection peaks below them according to the 8 foot and 18 foot A0 spacings.

Figure 4-16 is a 1940 Lane Wells ES. There is an SP (Natural Potential), a short normal (curve 2), two laterals (curves 3 and 4), and a point electrode traced on the log. The point electrode shows more variations than the short normal as it is more influenced by the borehole diameter and thin beds. From this log, it is almost impossible to determine the lateral curve spacings although they appear to be right side up. The lower boundary of the resistive zone (in the oil sand) appears 16 feet from the bottom (from the laterals). This would make me believe, if the laterals are on depth, that there is a possibility of water in the bottom of this zone.

More log examples will be shown and discussed in the chapter on the Microlog and the chapter on pulsed neutron capture and ES logs.

Figure 4-17 shows a 1940's ES by Lane Wells. Careful examination will show that the lateral is inverted from the normal lateral we have discussed. This means that the decay zone on the lateral is at the

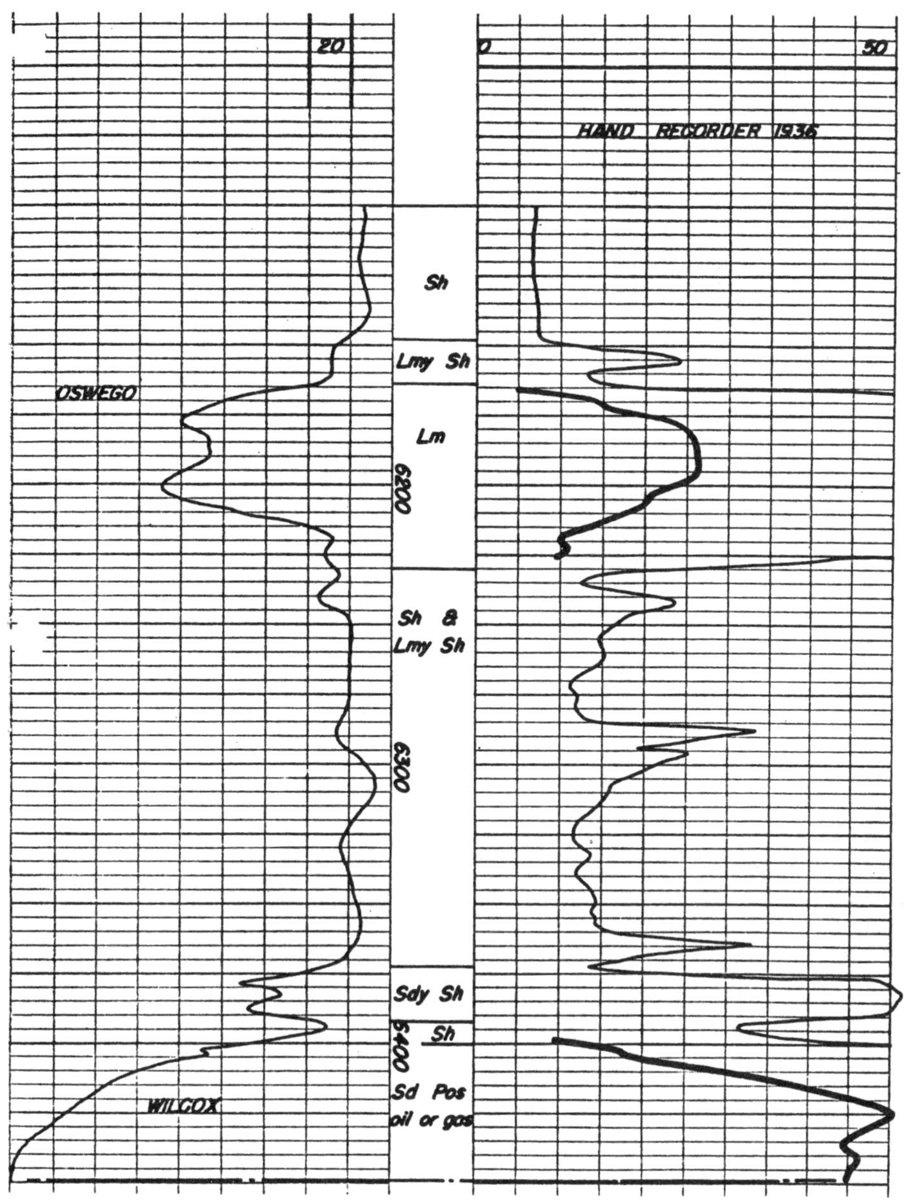

Figure 4-9. An old ES from the Oklahoma City Field (Wilcox Oil)

bottom of the bed and the peak is at the top. Picking the oil-water contact is thus very difficult as the water zone will merge into the decay zone on the lateral. On this log I would pick the water-oil contact at least 4 feet lower than is indicated on the figure. I think the oil-water contact is somewhere between 3200 and the bottom of the bed. In fact, it could be as low as 3209.

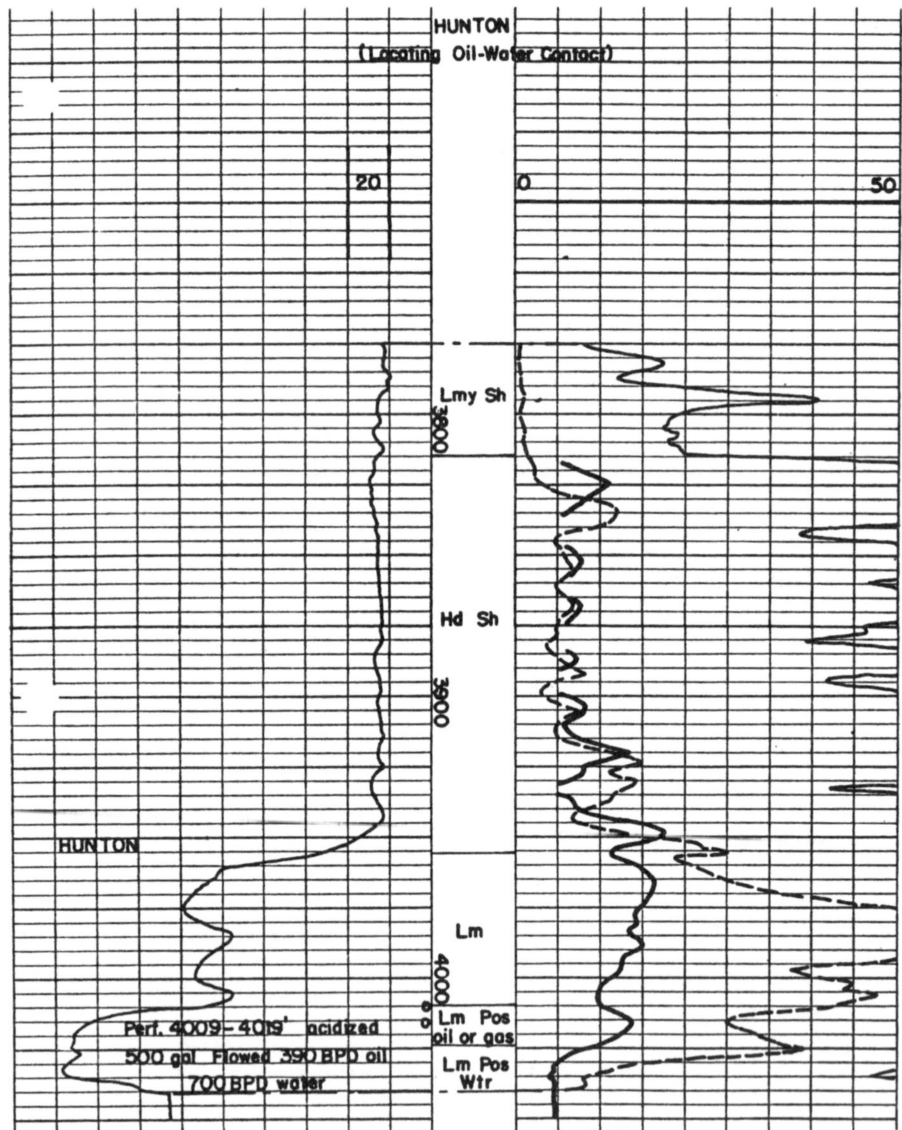

Figure 4-10. An Old ES Showing and Oil Water Contact

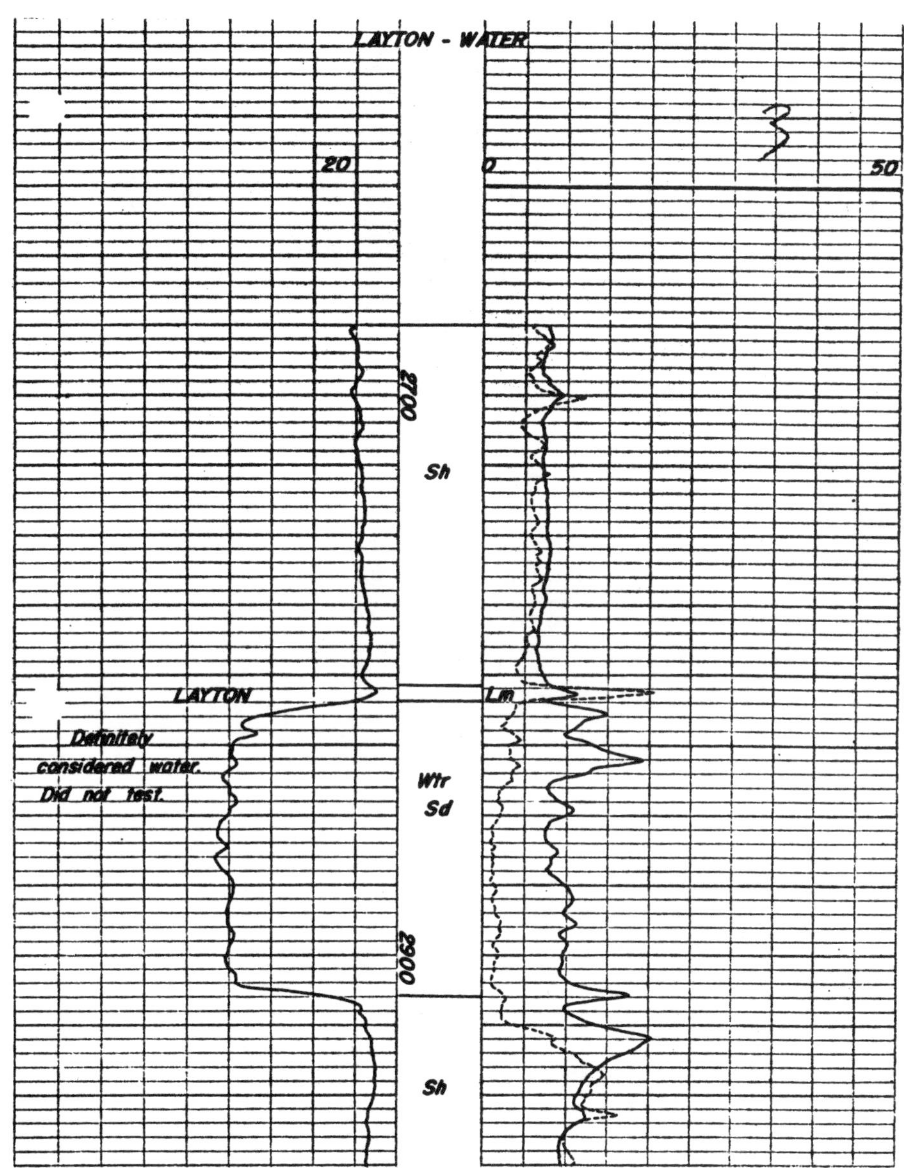

Figure 4-11. An Old ES Showing a Water Zone in Oklahoma

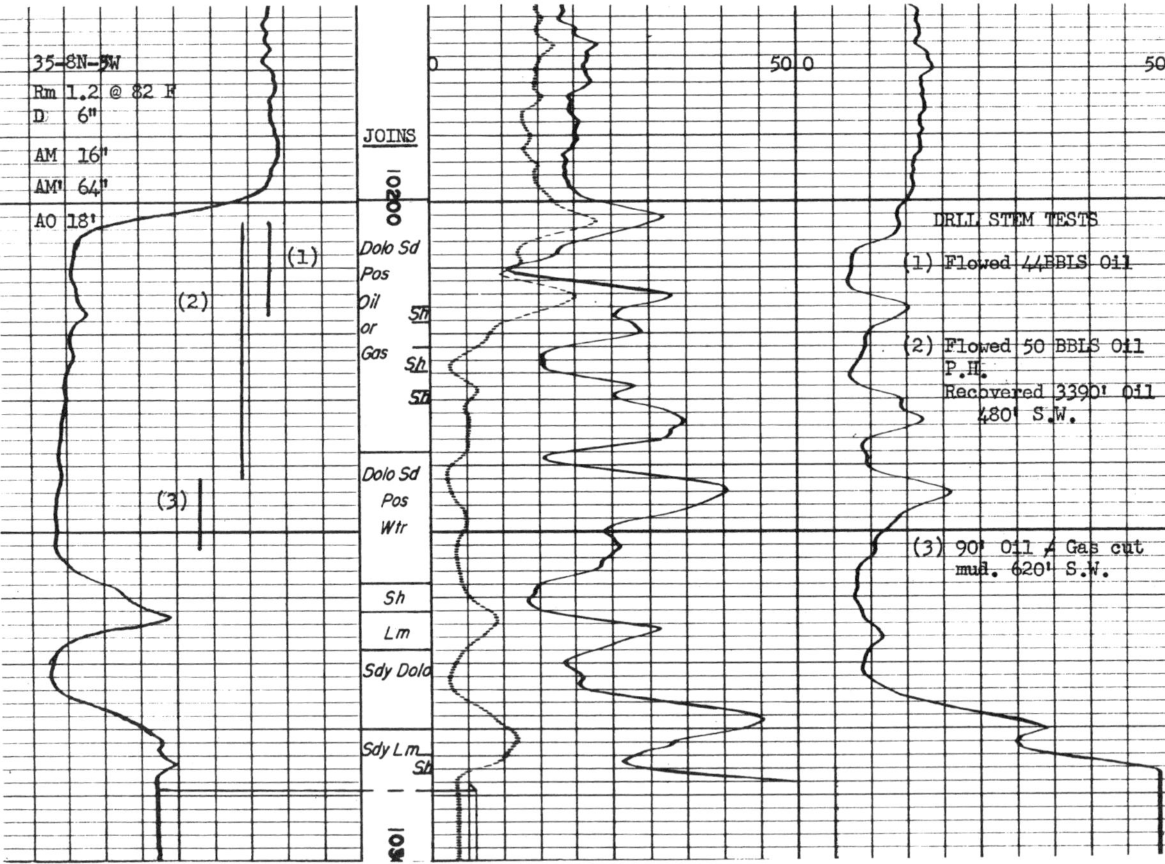

Figure 4-12. A Modern Oklahoma ES Showing an Oil-Water Contact.

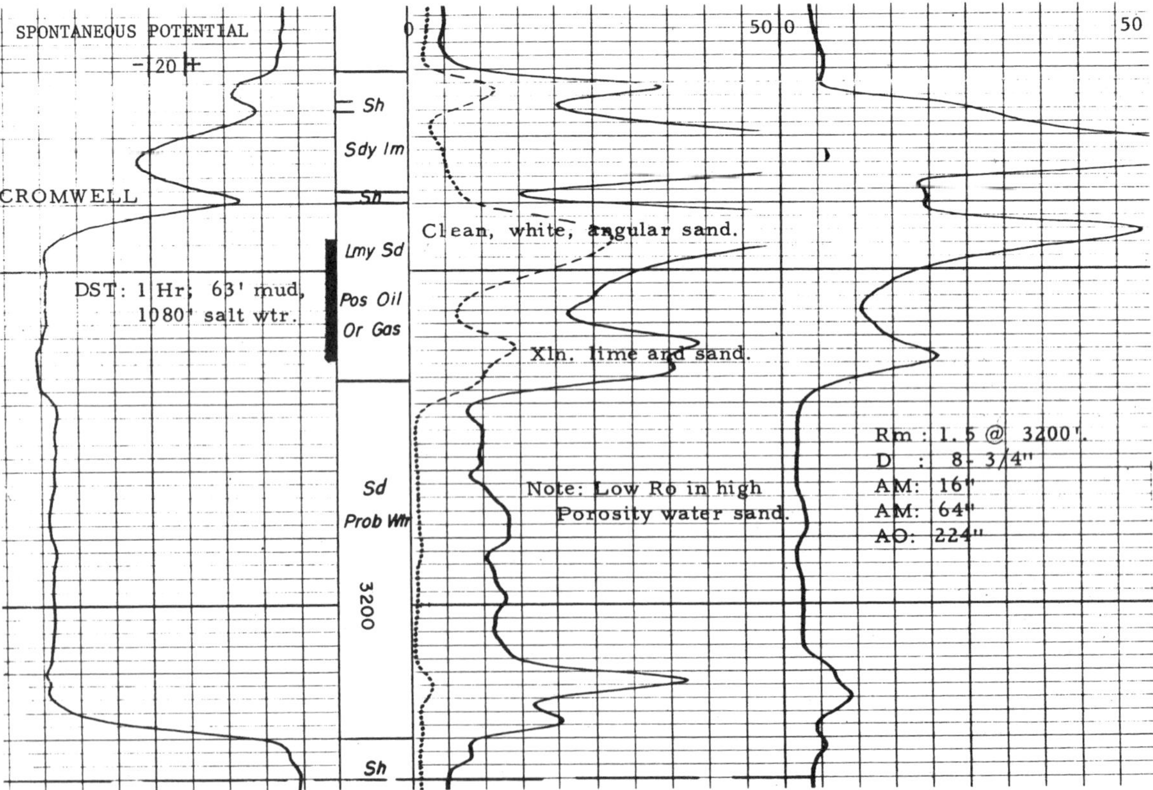

Figure 4-13. A Modern ES Showing a Water Bearing Zone with a Decrease in Porosity at the Top

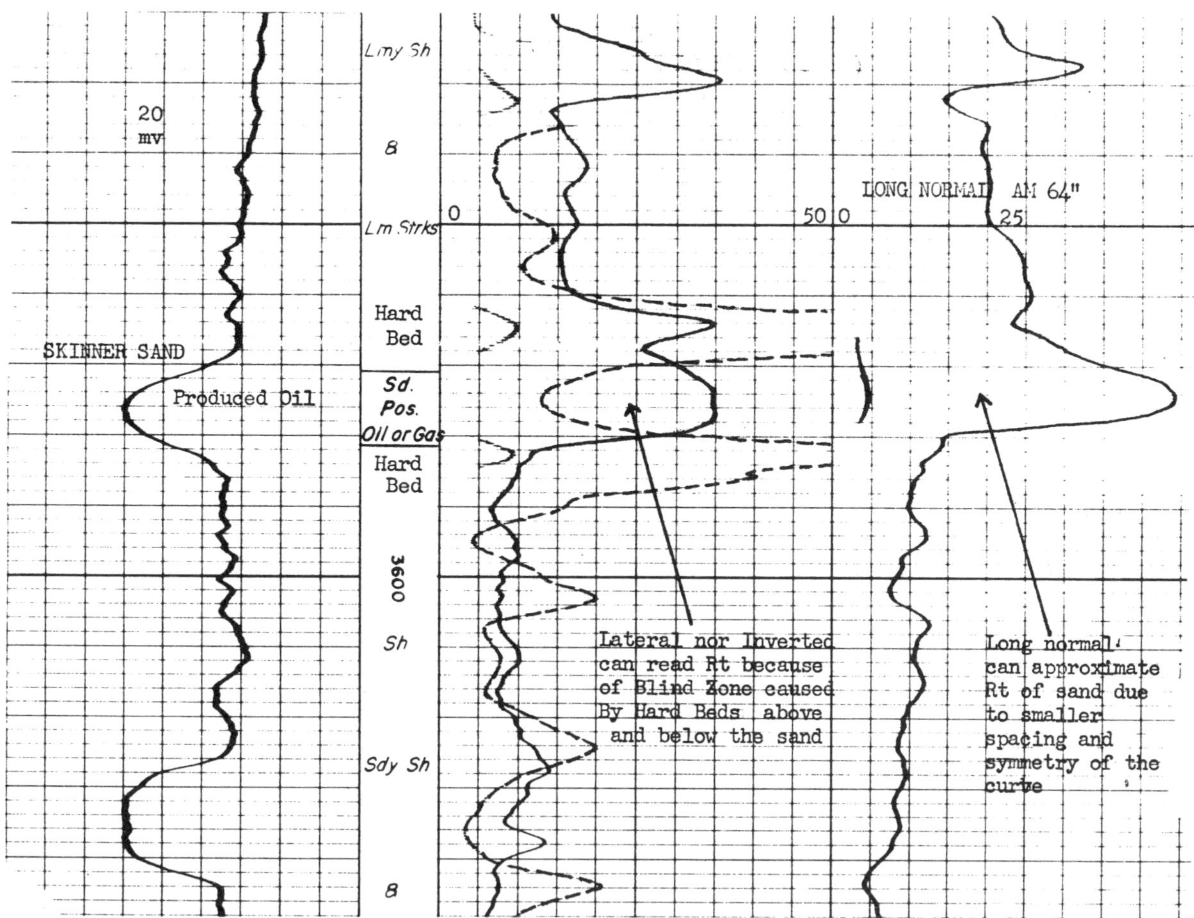

Figure 4-14. A Modern ES Showing a Thin Oil and Water Zone Comparison

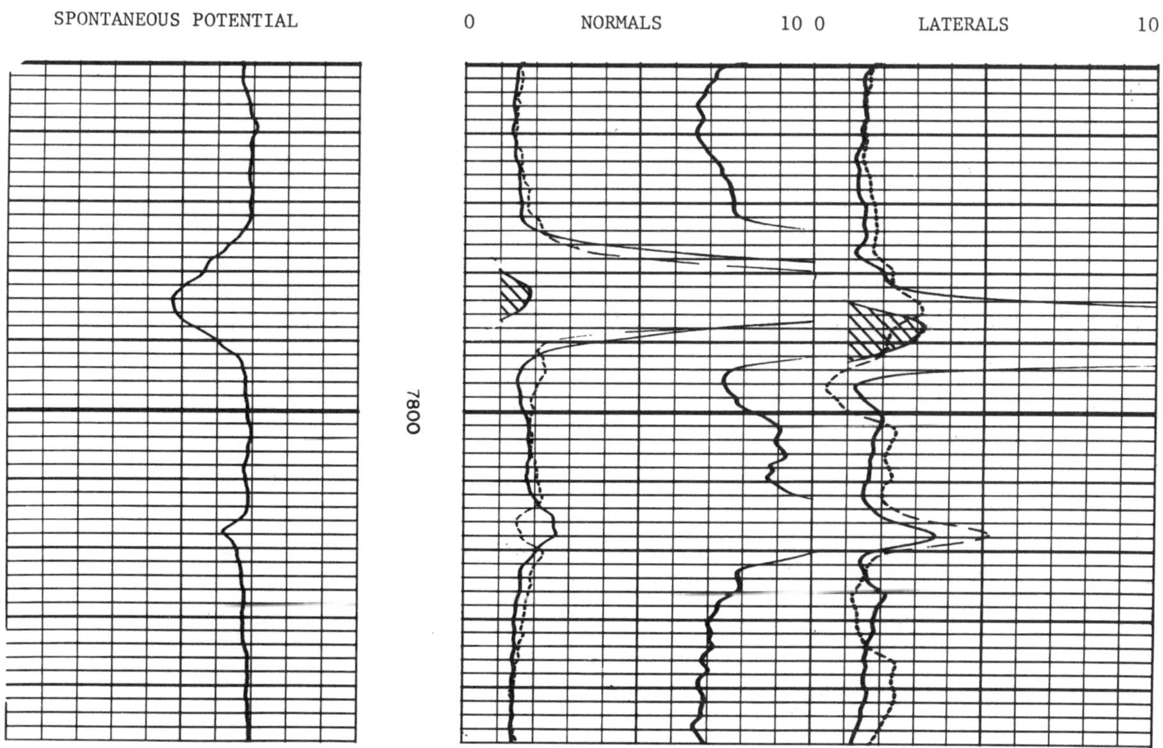

Figure 4-15. A Modern PGAC ES with Four Resistivity Curves

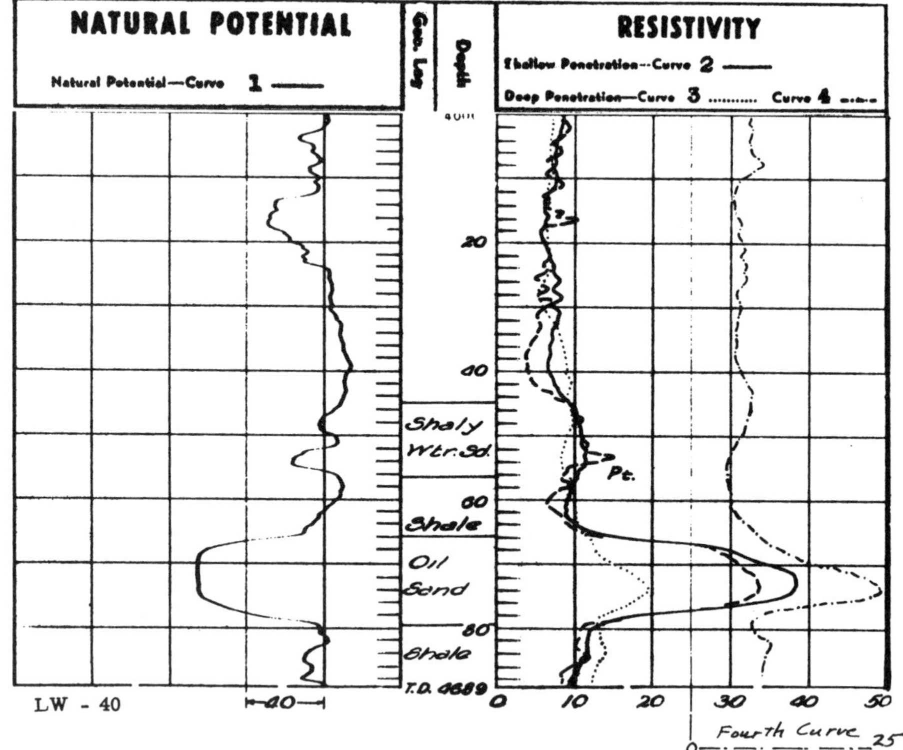

Figure 4-16. A 1940's Lane Wells ES with a Point Electrode Curve

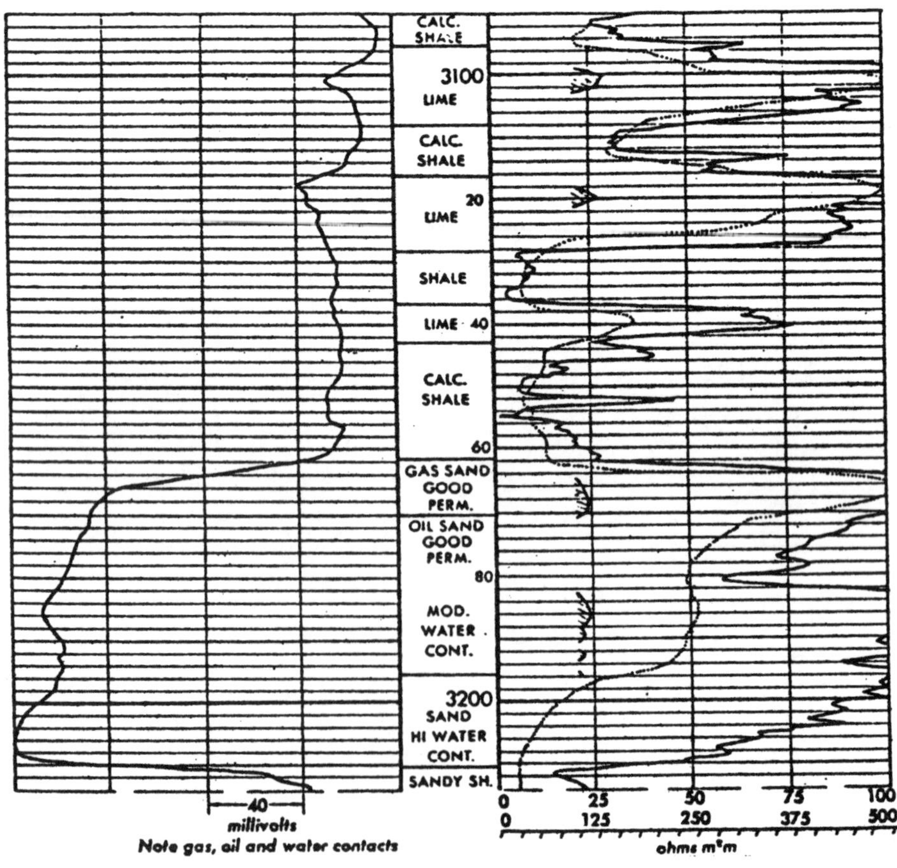

Figure 4-17. A 1940's Lane Wells ES

5

Hilchie, D. W., Electrical log departure curves, Old electrical log interpretation, p. 49–59.

Electrical Log Departure Curves

Electrical Log Departure Curves, as discussed in this chapter, are used to correct the ES readings for invasion effects to obtain Rt, and in some cases Ri, or to correct ES readings for invasion and bed thickness influences to obtain Rt and Ri. To correct for invasion, you need to have three or more resistivity curves with different depths of investigation. The three unknowns are Rt, Ri and Di (diameter of invasion).

HISTORICAL

The first set of departure curves were published by Schlumberger as Document 1 in 1947. This was for infinite thickness beds only and used a curve fitting technique. This was followed by Document 3 in 1949 for beds of finite thickness and later Document 7 in 1955 for infinitely thick beds of higher resistivity (an update of Document 1). These all used curve fitting to obtain the answer. Di/d (diameter of invasion to borehole diameter) up to 10 could be handled. These documents are presently out of print and have been for many years.

Welex introduced departure curves for infinitely thick beds for their two lateral and normal combination of curves. To use simplified departure curves, it is necessary to correct the curves for bed thickness influences when the beds are thin. These were curve fitting systems also.

PGAC also introduced a set of simplified departure curves that corrected for bed thickness, Di/d up to 5, for their three normal Mid-Continent log system (AM's, of 16, 64 and 38 inches). These were only usable for the PGAC Mid-Continent logging system.

Lane Wells in 1956 introduced a set of departure curves based on the original work of L. de Witte (*Oil and Gas Jour.* Feb. 11 1957, page 117) for their and Schlumberger's 16 inch and 64 inch normals combined with an 18 foot, 8 inch lateral. This system does not require curve fitting. This is the system that will be used in this book. This set of curves has been out of print for many years also.

In 1959 Guyod and Pranglin introduced a set of departure curves based on their analog model work. This is the best and most accurate set of departure curves available to the industry. Bed thickness and Di/d of up to 10 are made with these departure curves. These were designed for 16 and 64 inch normals combined with an 18 foot 8 inch normal. This set of curves is used much like the Lane Wells charts but is more complex. The serious student of old ES logs should have a set of these charts. These charts are presently available from:

Hubert Guyod
P.O.Box 36026
Houston, Texas 77036

DEPARTURE CURVES

Lane Wells departure curves presented in this chapter correct only for invasion effects. Corrections for bed thickness effects are found in chapters 2 and 3. These use the most commonly available curve spacings, 16 and 64 inch normals combined with 18 foot, 8 inch lateral. Curves obtained with other spacings and combinations can be solved for with the Schlumberger charts. The charts presented here were chosen for their simplicity, ease of use and limited number of charts. Chart sets like Guyod and Schlumberger represent from 120 to over 200 pages of charts. This is too many for any but the most serious students.

The charts presented are for 7 7/8 inch diameter borehole. These may be used for other borehole diameters but the results are not quite as accurate. Four sets of charts are presented for Di/d of 2, 5, 10 and 15. Each set contains a chart for the 16 inch normal versus the 64 inch normal and the 16 inch normal versus the 18′8″ lateral.

To use the charts you obtain the resistivity from the 16″, 64″ and 18′8″ curves and correct for bed thickness effects using Figures 2-16, 3-9 and 3-10. Remember for Figure 2-16 use the peak resistivity. If you do not make these corrections, use the raw data for the normals and Rat (Figure 2-12) for the lateral. Divide each of these three values by the mud resistivity (Rm) for the depth of the formation. You are now ready to enter the Departure Curves (Figures 5-1 to 5-8).

NORMAL
LANE WELLS
A Division of Dresser Industries, Inc.
$d = 7\frac{7}{8}''$
$\frac{D_i}{d} = 2$
$\frac{R_t}{R_m} =$ ◇
$\frac{R_i}{R_m} =$ ▭
Figure 5-1
$\frac{R_{16''}}{R_m}$
$\frac{R_{64''}}{R_m}$

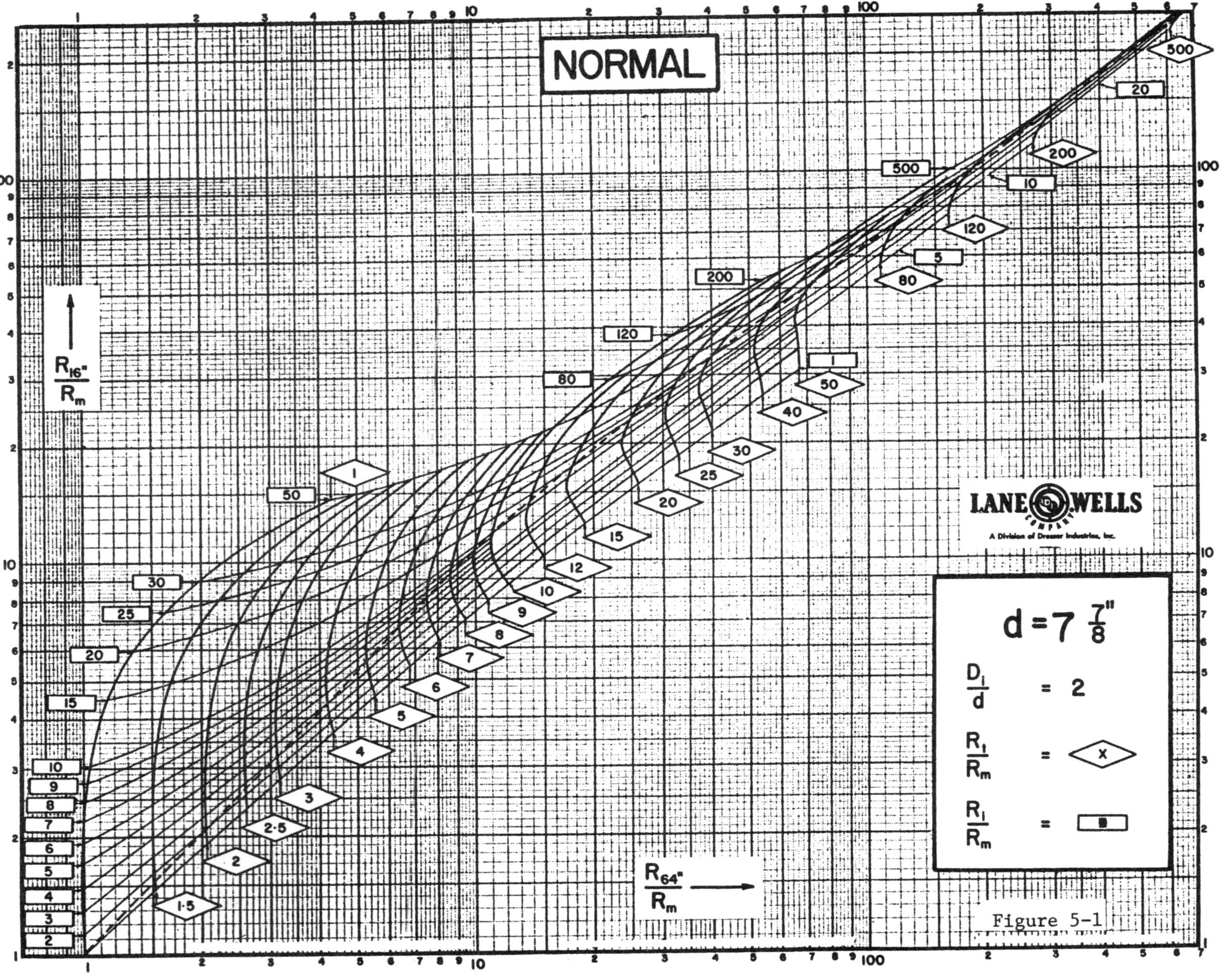

Figure 5-1.

LATERAL
LANE-WELLS
A Division of Dresser Industries, Inc.
d=7 7/8"
D_i/d = 2
R_t/R_m =
R_i/R_m =
Figure 5-2
R_18'8"/R_m
R_16"/R_m

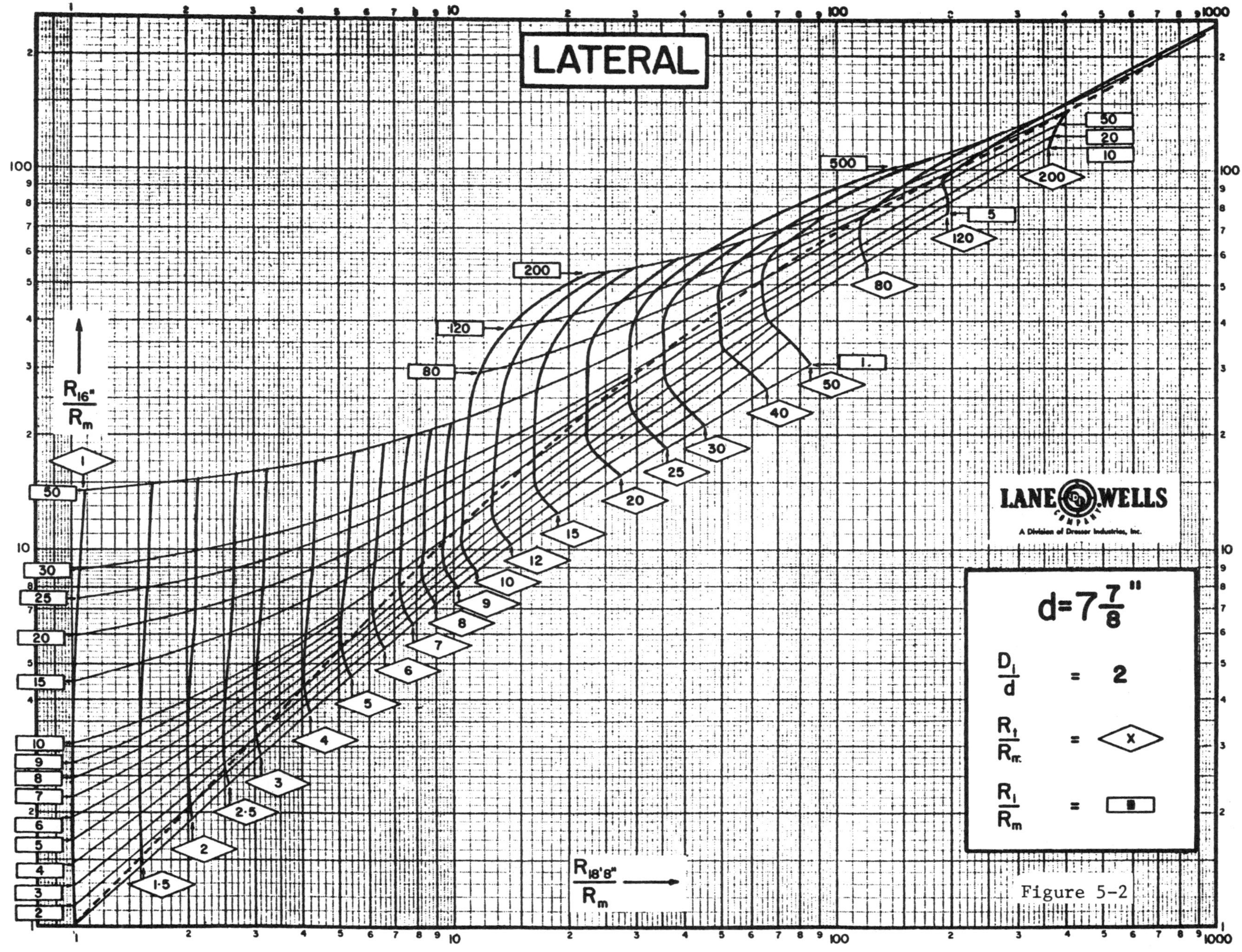

Figure 5-2.

NORMAL
$\frac{R_{16"}}{R_m}$
$\frac{R_{64"}}{R_m}$
LANE WELLS
A Division of Dresser Industries, Inc.
d = 7 7/8"
$\frac{D_i}{d}$ = 5
$\frac{R_t}{R_m}$ = ◇X◇
$\frac{R_i}{R_m}$ = ▭
Figure 5-3

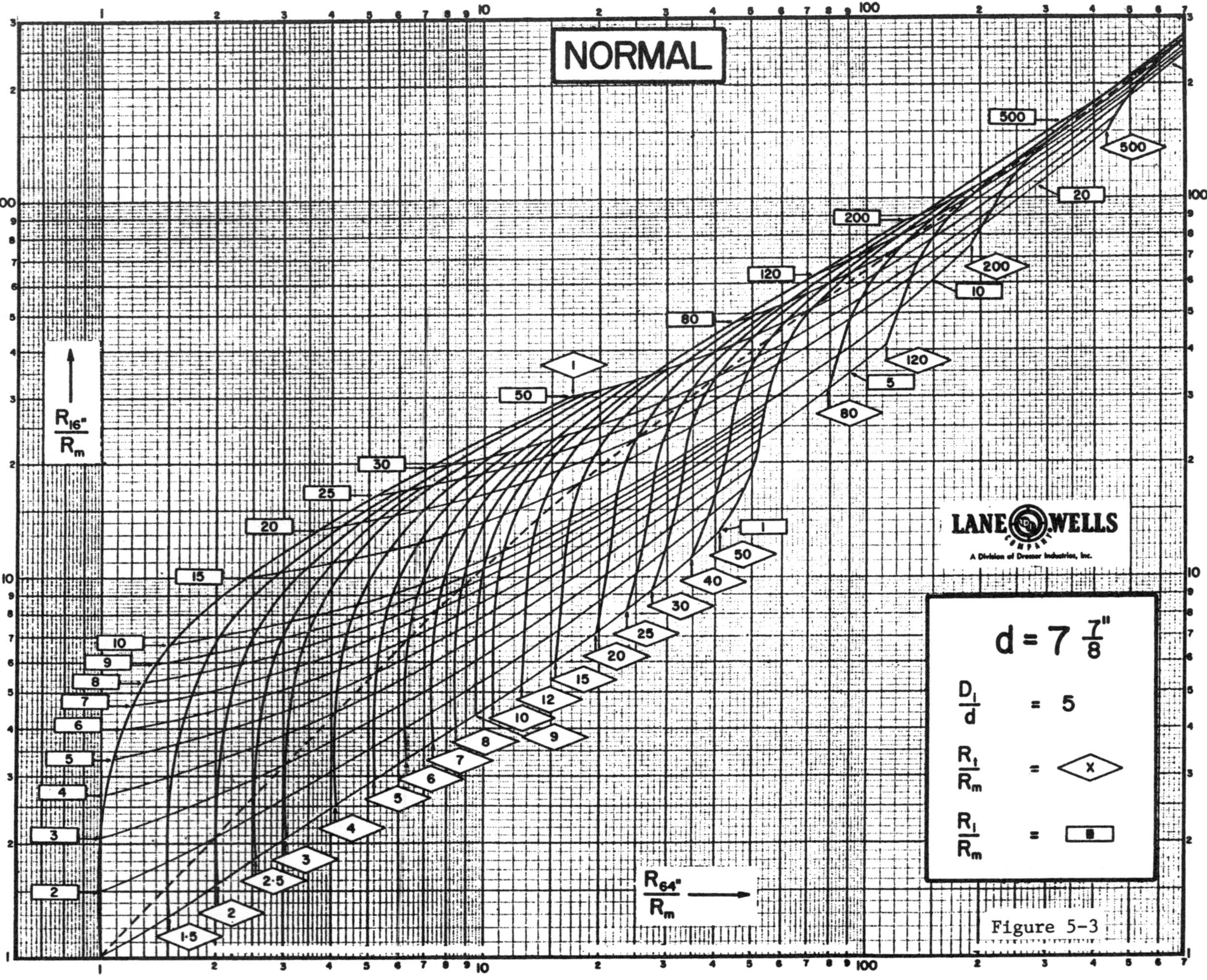

Figure 5-3.

LATERAL
$R_{16"}/R_m$
$R_{18'8"}/R_m$
LANE WELLS COMPANY
A Division of Dresser Industries, Inc.
$d = 7\frac{7}{8}"$
D_i/d = 5
R_t/R_m = ◇
R_i/R_m = ▭
Figure 5-4

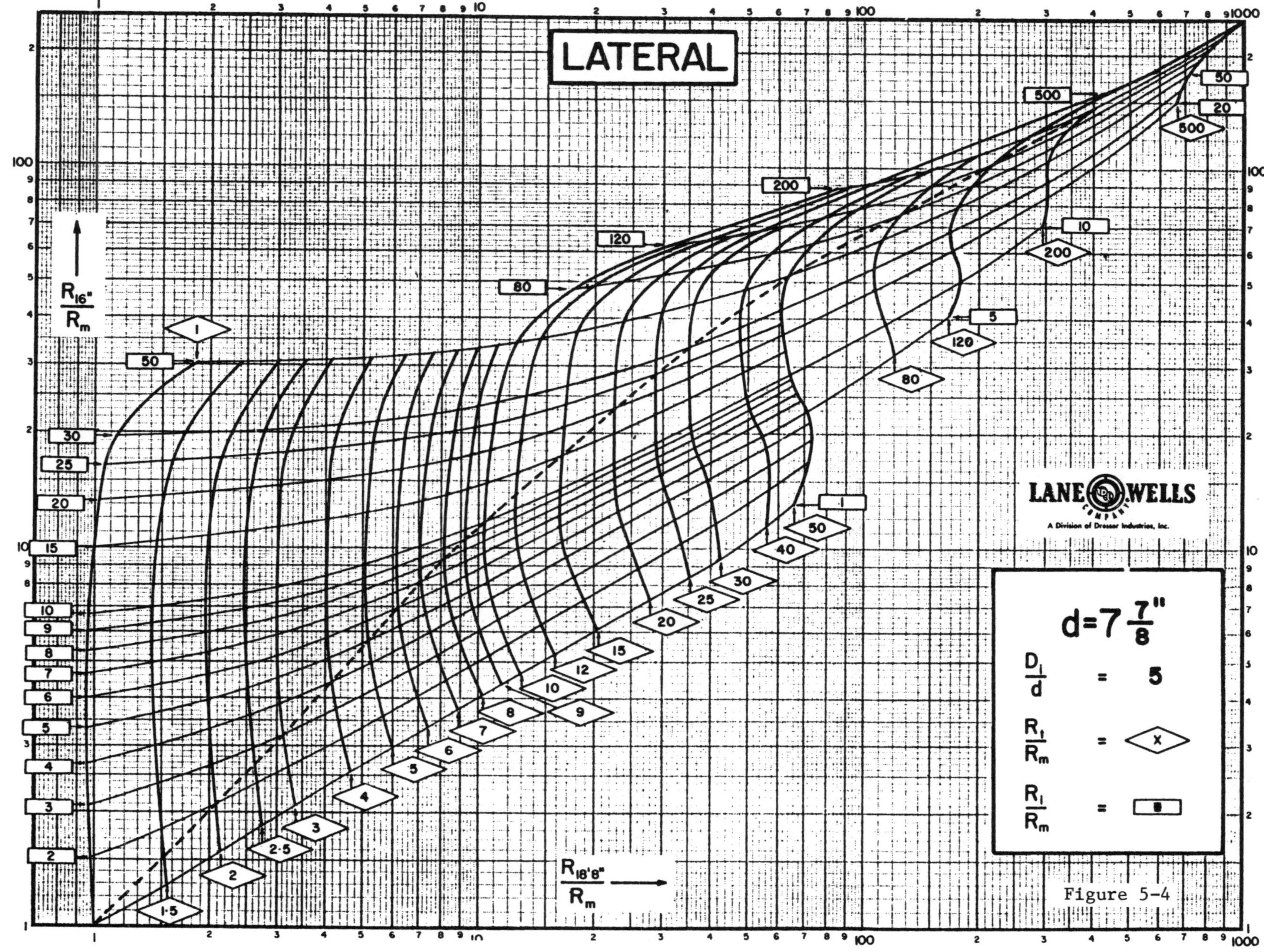

Figure 5-4.

NORMAL
LANE WELLS
A Division of Dresser Industries, Inc.
d = 7 7/8"
Di/d = 10
Rt/Rm = ◇
Ri/Rm = ▭
R64"/Rm
R16"/Rm
Figure 5-5

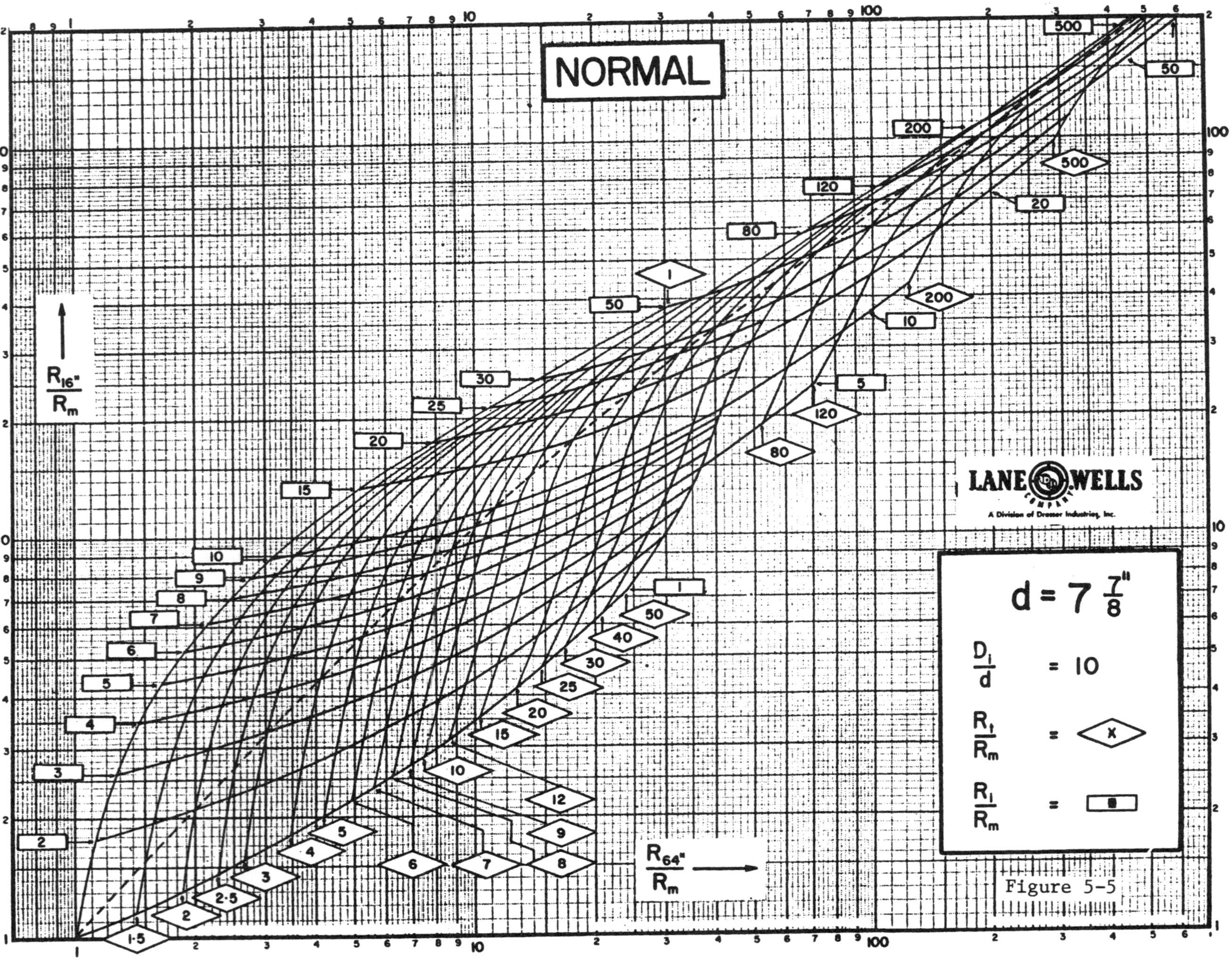

Figure 5-5.

LATERAL
LANE-WELLS
A Division of Dresser Industries, Inc.
d=7 7/8"
$\frac{D_i}{d}$ = 10
$\frac{R_t}{R_m}$ = ◇
$\frac{R_i}{R_m}$ = ▭
Figure 5-6
$\frac{R_{16"}}{R_m}$
$\frac{R_{18'8"}}{R_m}$

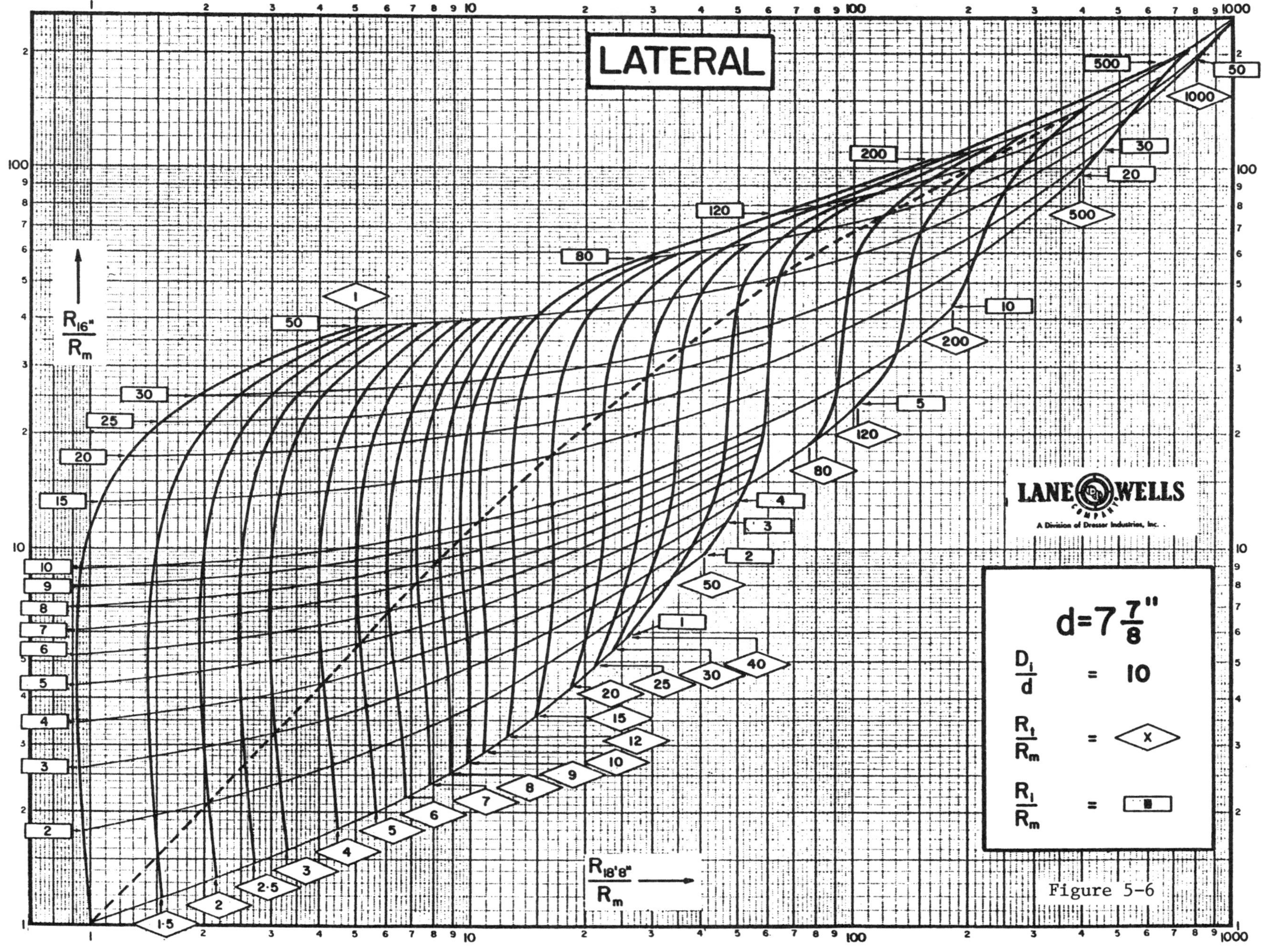

Figure 5-6.

NORMAL
LANE-WELLS
A Division of Dresser Industries, Inc.
d = 7 7/8"
D_i/d = 15
R_t/R_m = ◇
R_i/R_m = ▭
$R_{16''}/R_m$
$R_{64''}/R_m$
Figure 5-7

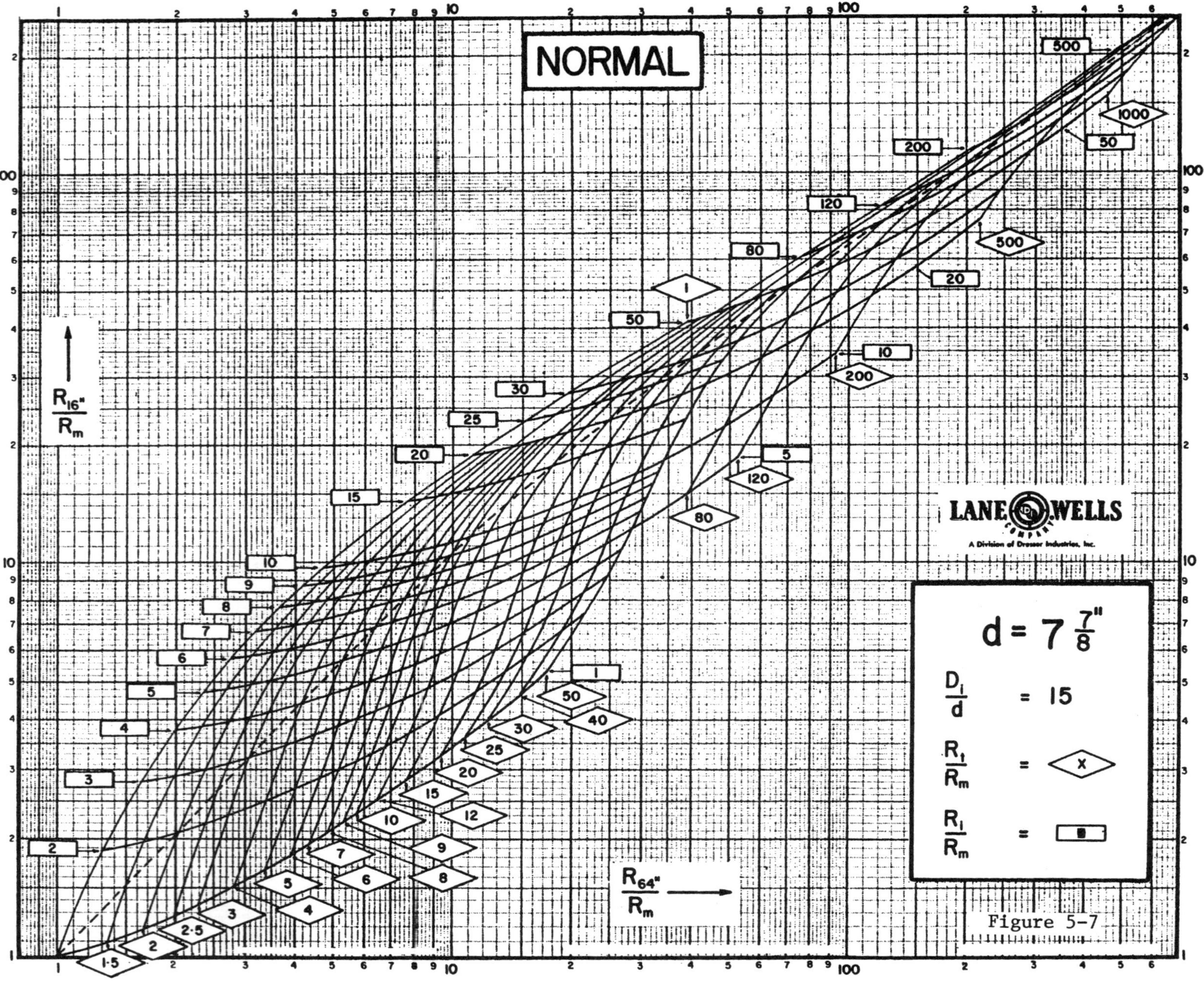

Figure 5-7.

LATERAL
LANE WELLS COMPANY
A Division of Dresser Industries, Inc.
d=7 7/8"
D_i/d = 15
R_i/R_m = ◇x◇
R_i/R_m = ▭
Figure 5-8
R16"/Rm
R18'8"/Rm

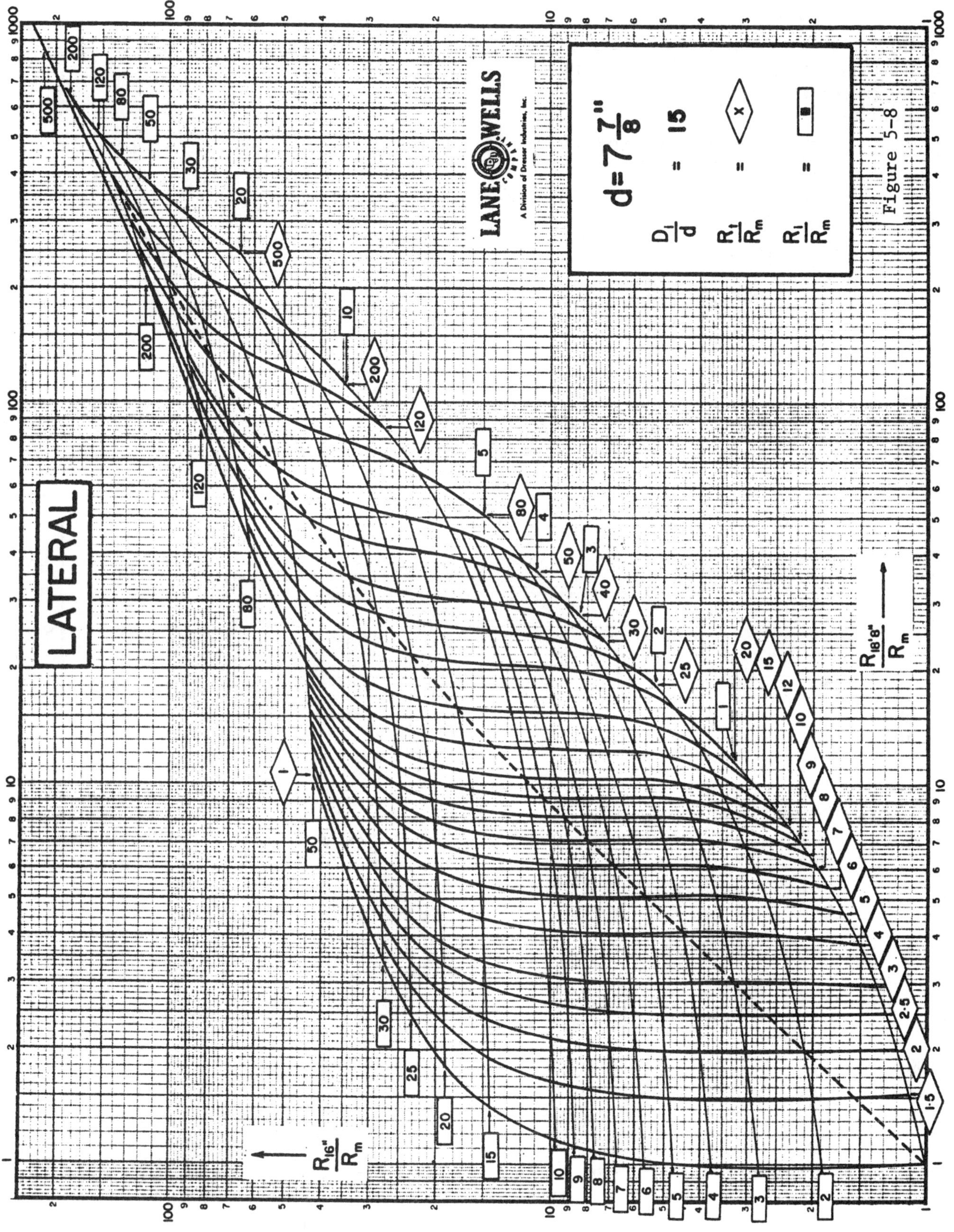

Figure 5-8.

From each chart (Figures 5-1 to 5-8), obtain a value of Rt/Rm. On a piece of log-log graph paper, plot Rt/Rm versus Di/d for the two normal charts (Figures 5-1, 3, 5 and 7). Do the same thing for the 16″ normal versus lateral data (Figures 5-2, 4, 6 and 8). Where the two curves that are drawn through the plotted data points cross, read Rt/Rm. This is the solution.

Example 5-1 shows the determination of Rt for the log shown in Example 5-1.

DISCUSSION

The three resistivity curves must be reading resistivity relatively accurately or the drawn curves may not cross (see part 2 of example 5-1). If invasion is very shallow the curves may not cross but may be converging at low values of Di/d. This is typical of most of the Gulf Coast logs I have worked. If invasion is deep, the drawn curves may be converging beyond the range of the charts. If you do not having crossing of the drawn curves, you may estimate or guess Rt based of the data or assume that the lateral reads Rt. If Rm is too badly off, it may cause problems. Sometimes by changing Rm you can cause convergence. This correcting of Rm may not result in a unique solution and this should be checked by several different Rm's. This is very tedious when done by hand.

If the invasion front reaches beyond the medium investigation curve, the data will not converge and no solution will be obtained. If the resistivity curves do not stack properly, after bed thickness correction, the drawn curves will not give you an Rt/Rm convergence.

EXAMPLE 5-1

Log used see Example 5-1 Log A.

The Rm for this log is .7 at 2800 feet. The bed thickness is a little hard to pick. The overall resistive bed looks like about 33 feet to me, but one could easily justify a thinner pick. One of the problems with ES surveys is that you have to pick 1 resistivity value for the bed even if it is not constant. Picking the equivalent bed thickness is usually even harder. This is one of the reason many analysts do not make correction to the curve readings. Rs from the log looks like 5 ohm m.

	$R_{from\ log}$	$R_{corrected}$	$R_{corr.}/Rm$
R_{SN}	28	28	40
R_{LN}	23	26	37
R_L	21 (peak)	16	23

From Figures 5-1 to 5-8 and in-putting the above data:

Di/d	*Rt/Rm (2 normals)*	*Rt/Rm (SN & lateral)*
2	20	18
5	15	19
10	10	18
15	1	15

Plotting this data you get:

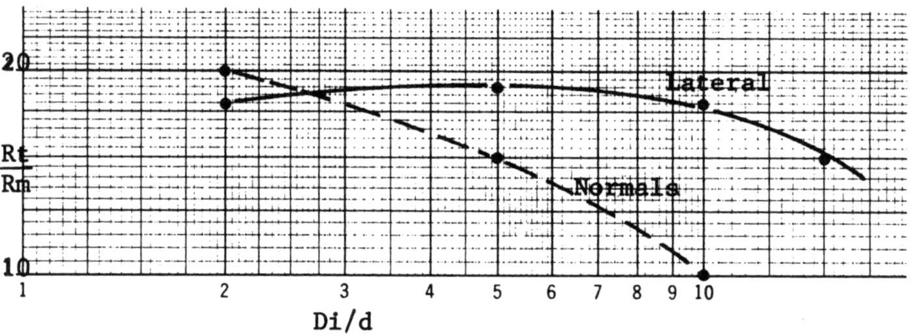

Rt/Rm from where the curves cross is 18.5.
Rt = Rt/Rm × Rm = 18.5 × .7=13 ohm m.

Log B run 14 days after Log A

		R_{corr} *(e = 33 ft)*
R_{SN}	40 ohm m	40
R_{LN}	60 ohm m	70
R_L	55 ohm m	39
R_{LN}	appears to be reading too high - data does not stack logically. The Rt/Rm lines do not converge on plot. On Log B, the long normal in the shales 50 feet above (not on log) read 5 ohm m, while the short normal and lateral read 4 ohm m. The long normal is out of calibration. This is about the amount of error, percentage wise, in the long normal reading.	

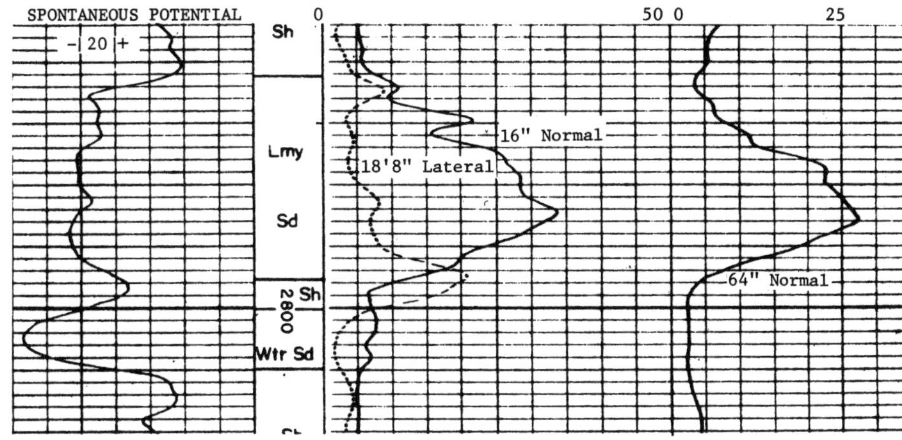

Example 5-1. Log A

This is a water bearing zone. Accordingly, the first log(A) appears to be right qualitatively and quantitatively. Log B then has invasion much deeper than Di/d of 15 which is the limit of the departure curves.

More examples will be worked later in the book.

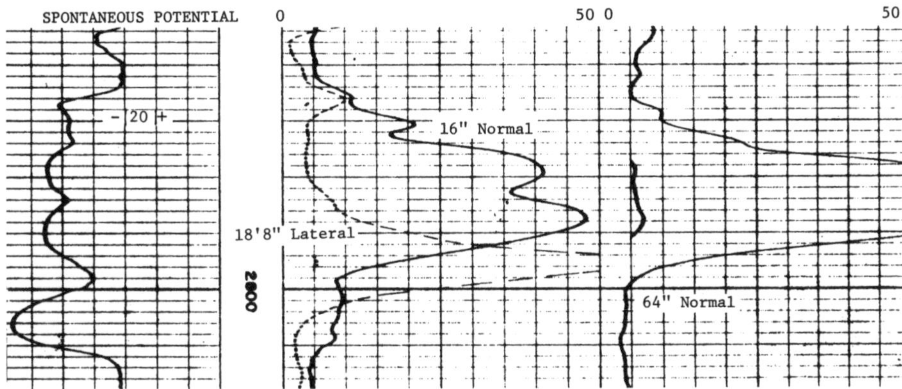

Example 5-1. Log B Run 14 Days after Log A. Mud stayed the same.

Hilchie, D. W., The micrSolog, Old electrical log interpretation, p. 61–70.

6

The Micrslog

The Micrslog was commercially introduced in 1948. Although the term Micrslog is a trade name, it is used generically in the oil business. The same type of measurement is called a Contact Log by Halliburton, Micro-contact by Birdwell, Micro-survey by PGAC, a Permalog by Elgen and a Minilog by Lane Wells.

THE MEASUREMENT

The Micrslog is a pad resistivity measurement. The pad with three electrodes (as shown in Figure 6-1) is forced against the side of the hole. The sonde pictured in Figure 6-1 is a modern power-operated device in which the arms can be retracted and extended by surface controls. The pad is hydraulic. That is, it is filled with oil to create a better contact with the formation. The older and original pads were solid rubber. The older pads were held against the side of the borehole by a bow spring. The older models, which were used until the mid 1950's, were run in the hole closed and opened by firing a plug when the total depth was reached. The tool then stayed open all the time it was in the borehole.

Two resistivity measurements are made by the three electrodes on the pad which are one inch apart. A 2 inch normal is run using the lower electrode as the current electrode and the upper electrode as a potential measurement electrode. A 1 inch by 1 inch inverse (1″ × 1″ or $1^1/_2$″) is run using a lateral type arrangement with the lower electrode being the current electrode and the two upper electrodes being the differential potential measurement. Figure 6-2 shows the two systems. The 2 inch "Micronormal" has a depth of investigation of about 4 inches while the 1″ × 1″ "Microinverse" has a depth investigation of about $1^1/_2$ inches. These depths of investigation, under the logging conditions, are a little too deep but sufficient to get the idea across.

When the pad is against a formation directly, both curves read the formation resistivity. If the formation is low porosity and impermeable, the resistivity will be very high. If the formation is shale, the resistivity will be low. If the formation is permeable, the mud will filtrate into the formation due to the higher borehole pressure. This filtration process results in the mud solids forming a cake (mudcake) on the side of the borehole and the filtrate (mud filtrate) entering the permeable formation. In a fresh mud system the mud cake is usually tough and resilient. The resistivity of this mud cake is something like 10 to 20% higher than the mud resistivity. The mud filtrate resistivity is often a little less than the mud resistivity. The Micrslog pad rests on this mud cake during the logging process. The Microinverse measures the resistivity of the mud cake and some of the formation. The formation is filled with mud filtrate and has a resistivity called Rxo (the resistivity of the

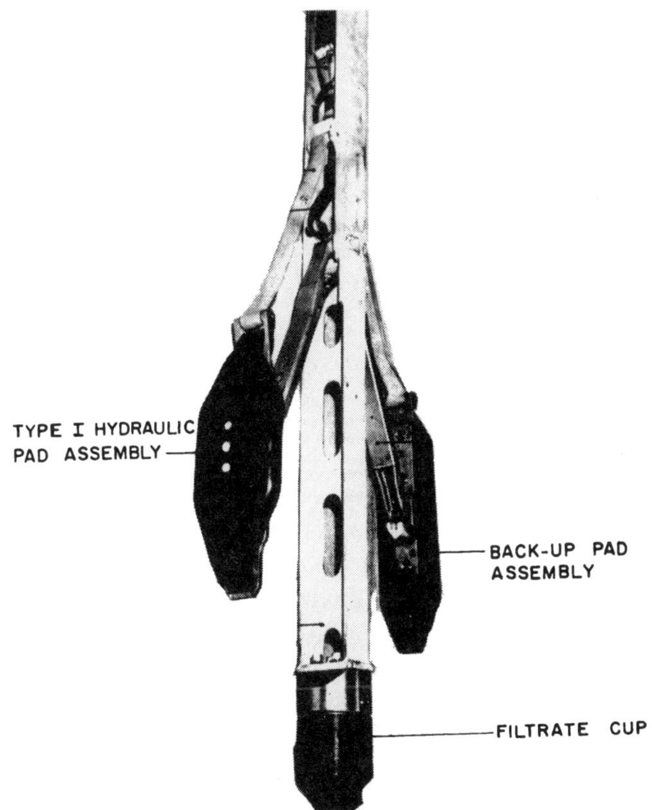

Figure 6-1. A Micrslog Downhole Sonde (courtesy Schlumberger)

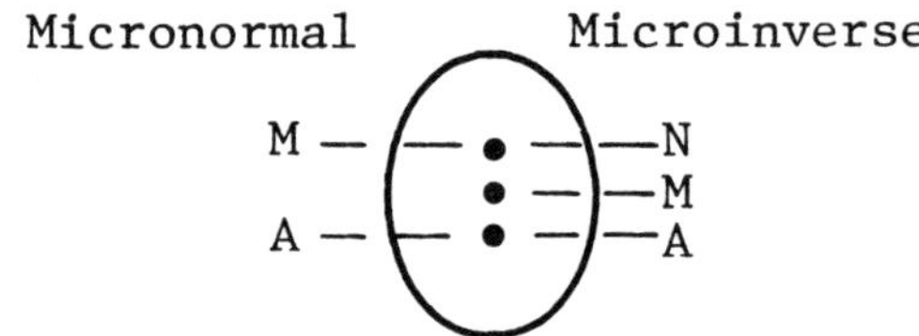

Figure 6-2. Microlog pad electrode array

flushed zone). The Micronormal, because of its larger depth of investigation, measures the resistivity of the mud cake and quite a bit of the formation (Rxo). As the formation filled with mud filtrate (Rmf) has a much higher resistivity, the Micronormal read higher than the Microinverse. This "positive separation" indicates a mudcake present which implies that the formation is permeable. The caliper (Microcaliper) run with the later model Microlog tools must be consulted to be sure the borehole is smooth and the positive separation is due to mud cake. If the borehole wall is rugose, positive separation can be created by the pad not being firmly against the formation.

Figure 6-3 is a simplified Microlog. The resistivity scales on Micrologs are limited to resistivities less than about 20 times Rm (mud resistivity). This is because permeable zone resistivities on the log occur around one tenth or less of this. Tite or impermeable zones show up as high resistivity. These may be recorded or may be off the log as the zone at the top of Figure 6-3. At these high resistivities, positive separation may occur but this should not be construed as permeablility. It is usually due to a poor fit of the pad against the formation. Often, opposite the permeable zones, the caliper will show borehole diameters less than bit size. This is due to the mud cake. This caliper mud cake determination does not always occur opposite permeable formations. Shales usually show up as low resistivity, as on the ES, with either no or negative separation (dotted curve with a lower resistivity than solid curve). In a severe borehole enlargement, the resistivity curves may read just mud. As the resistivity measured is a function of Rmc (mud cake resistivity), mud cake thickness and the formation resistivity next to the wellbore, the curve resistivities can be influenced by hydrocarbons. If the mud cake is relatively thin and the formation sufficiently porous, an increase in resistivity may occur as the tool moves from a water to a hydrocarbon bearing zone. The mud filtrate does not move all the hydrocarbons out of the flushed zone. These residual hydrocarbons (often called ROS for residual oil saturation) causes an increase in Rxo and the curves indicate this increase. If the invasion is not sufficient to create an Rxo (flushed) zone but a mud cake is created, positive separation may not occur. In a water zone the Rxo may be less than Rmc and no positive separation. Usually in hydrocarbon zones under these conditions, you get positive separation as the hydrocarbons supply the necessary resistivity contrast.

The Microlog is thus a mud cake or permeability detector. If the mud cake is not tough enough to hold the pad of the formation, the Microlog will not detect the mudcake. This happens in Gyp and Salt muds. If no positive separation is indicated at the bottom of the hole, it may be due to insufficient invasion.

The Microlog has excellent vertical resolution. It can detect mud cakes with only a few inches of vertical extent. This makes it particularly useful in carbonate where the permeablity is often very spasmodic. Where the permeability is broken and intermittent, the depth scale may be expanded. The Microlog may be presented on 10 inches or 25 inches per hundred feet of borehole.

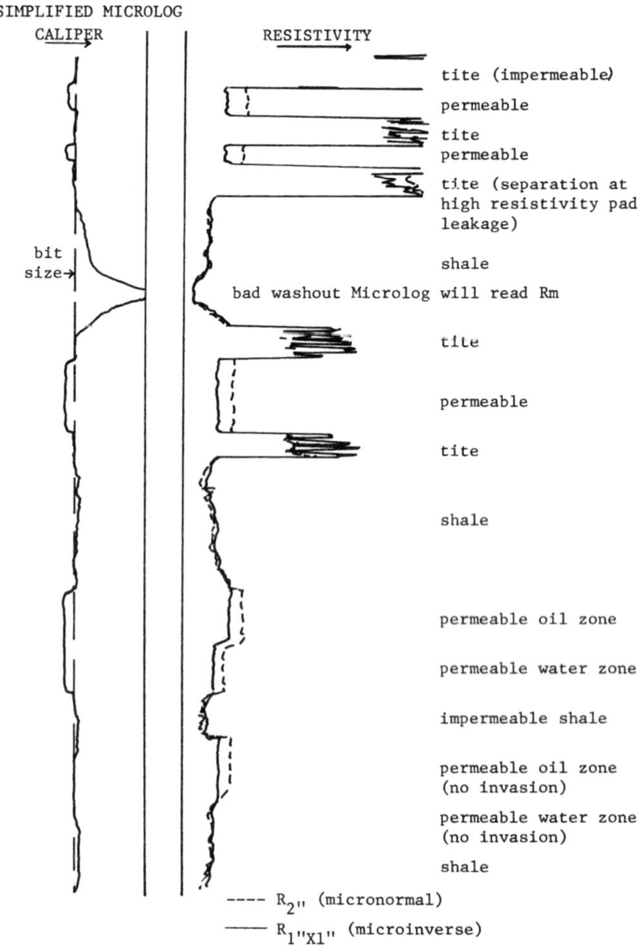

Figure 6-3.

On some Micrologs, a Micro-SP was run off an electrode on the backup shoe. This was not a very good SP and was only used for a short period of time.

POROSITY

The Microlog was initially considered a porosity tool. Today, the Microlog is not used as a porosity tool unless it is the only possible source of data. The philosophy is that if you can measure Rxo, you can then obtain porosity. The equations involved are:

$$F = \frac{.62}{\Phi^{2.15}} \tag{6-1}$$

for granular (sandstone) formations. F is formation resistivity factor and Φ is fractional porosity. Or:

$$F = \frac{1}{\Phi^{m}} \tag{6-2}$$

where m is the porosity (sometimes called the cementation) exponent. For carbonates when you don't know better m is equal to 2. So if you can determine F you can calculate porosity.

Rxo is related to F by:

$$F = \frac{\text{Rxo Sxo}^2}{\text{Rmf}} = \frac{\text{Rxo }(1 - \text{ROS})^2}{\text{Rmf}} \tag{6-3}$$

where Sxo is the saturation of the mud filtrate in the flushed zone, Rmf is the resistivity of the mud filtrate and ROS (is fractional residual hydrocarbon saturation) is 1 − Sxo.

To determine F from Rxo one must be able to estimate Sxo. Empirical relationships have been developed that seem adequate. These are:

if type of hydrocarbon is known

	API	*ROS%*	*Sxo%*
Gas		5–40	60–95
High Gravity	40–50	5–10	90–95
Low Gravity	10–20	20–30	70–80

Note ROS depends to some degree on Di

or if porosity range is known

Porosity%	*ROS%*	*Sxo%*
25–35	30	70
15–20	15	85

The best value of Sxo (or ROS) must be used as this represents a significant part of the porosity calculated. If an Sxo of 50% is used, you are doubling the porosity seen by the Microlog. If Sxo is 80%, you are only increasing the porosity by 25%. What this means is that if Sxo is low and you think it is high, you could miss the actual porosity by a significant amount.

The Microlog is usually recommended for porosities over 20%. Below this, the mud cake thickness influence becomes large and you don't know if the answer you calculate is close or not.

POROSITY CALCULATION

Figure 6-4 is the chart most service companies use for obtaining Rxo from the Micronormal and Microinverse readings from the log. The chart is entered with the readings from the Microlog divided by Rmc (resistivity of the mud cake). Rmc is not reliably measured in the field. In fact, the Rmc values measured are usually bad. The most reliable way to obtain Rmc is to go to a correlation such as Figure 6-5. These correlations require only a value of Rm (mud resistivity) at the formation temperature and the temperature to obtain Rmc (and Rmf if it is needed). As always, the mud sample used must be representative of what is in the hole. If the mud sample obtained by the logging crew is poor (such as from the pit), the results may be poor.

Entering Figure 6-4 with Micronormal and Microinverse resistivities divided by Rmc, you establish a point of the chart from which you can read Rxo/Rmc and mud cake thickness. If the mud cake thickness is known (which is very seldom), the Rxo/Rmc value can be corrected for mud cake thickness by moving the established point at 45° angle on the chart to the proper or established mud cake thickness. The mud cake thickness is that in front of the pad. If the caliper shows a mud cake contribution of 1/2 inch, the mud cake thickness in front of the pad is 1/2 of 1/2 inch or 1/4 inch. If the borehole diameter is something other than 8 inches, you multiply R for Microinverse by:

borehole diameter	*correction to* $R_{1'' \times 1''}$
4.75 inches	1.15
6	1.05
9	.975
10	.93

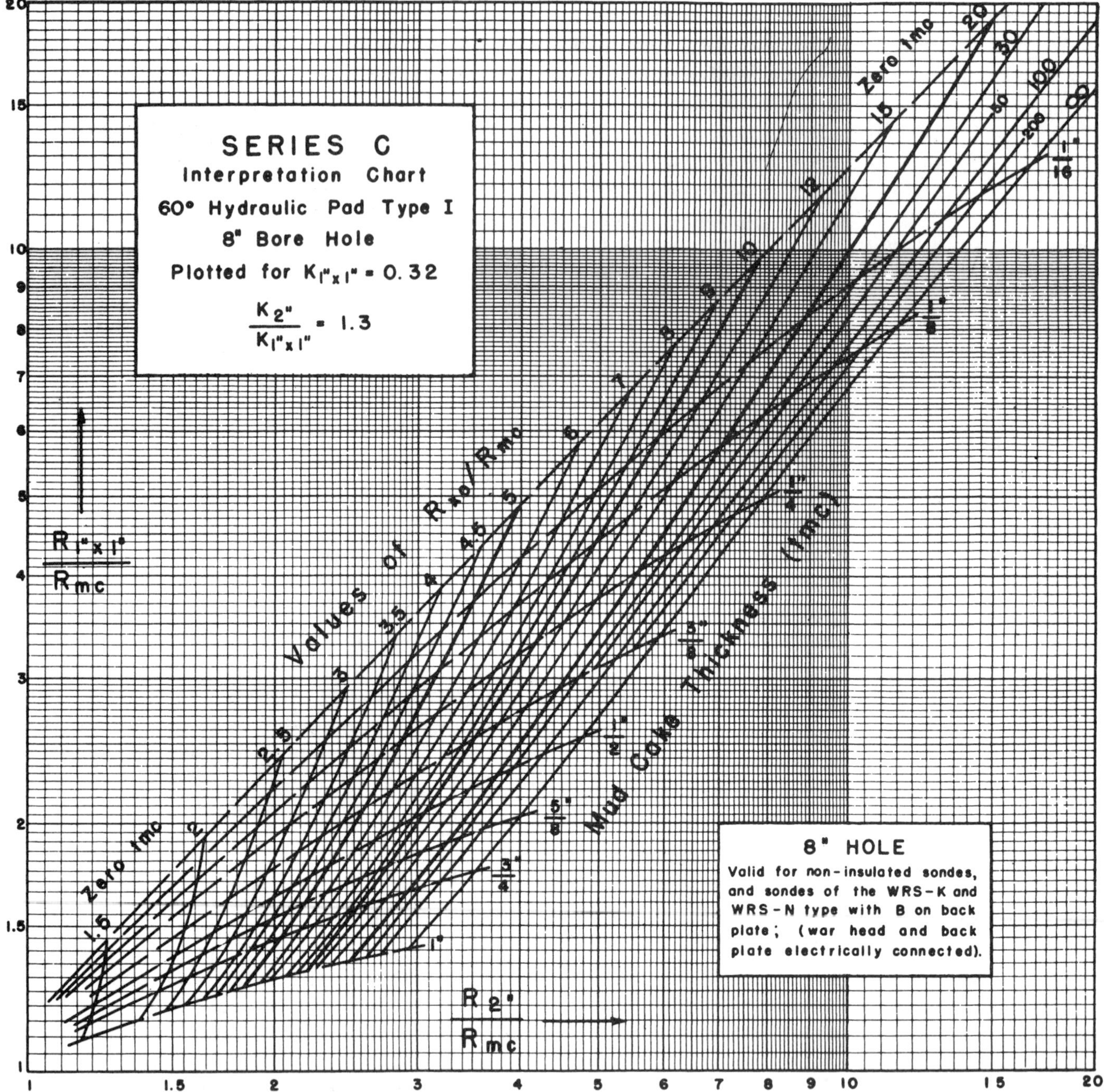

Figure 6-4. Rxo from Microlog (courtesy Schlumberger)

To go from Rxo/Rmc, you may use either equation 6-3 or Figure 6-6. In Figure 6-6 you input Rxo/Rmc and Rmc and calculate Rxo. You then input Rmf and obtain Fa. Fa is corrected using ROS and the result is porosity for a sandstone. This can be corrected for other m values with the rest of the chart. See Example 6-1 for more details.

If Rmc is not reliable, or the mud cake is thick or the porosities low, Figure 6-7 is often used. It works best for the "D" pad (solid) and gives porosities a little low for hydraulic pads.

If there are water sands available around the formation you wish to interpret, a simple way of checking how good an Fa you are getting is to crossplot Fa versus F from the water zones using the ES Ro's and the Rw for the formations. Good values should plot (on log - log graph paper) as a 45° line as shown in Figure 6-8. If the values fall above or below this

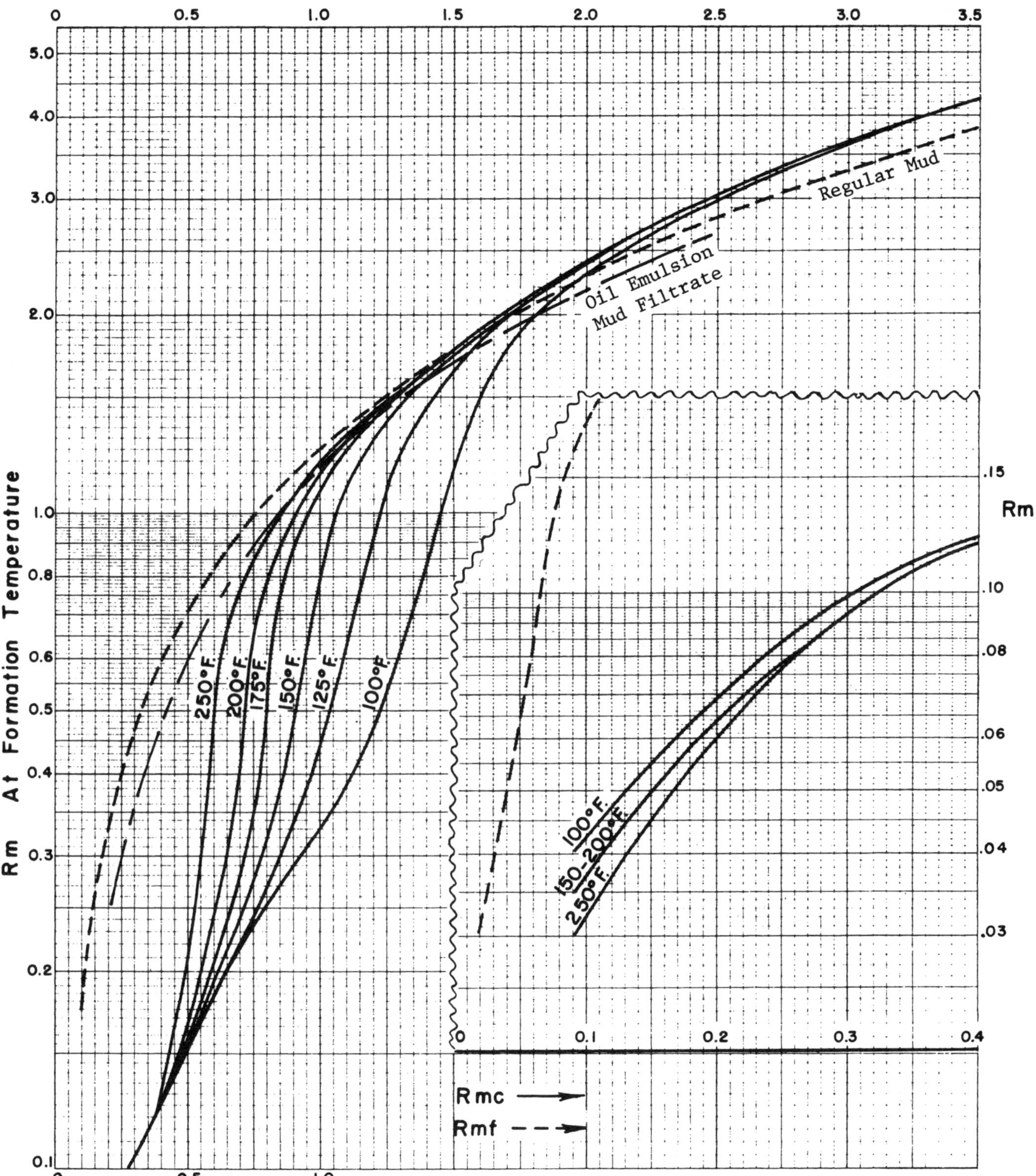

Figure 6-5. Rm - Rmc - Rmf Empirical Relationships

agreement line, the Fa values should be corrected to bring them onto the correct line.

Figure 6-8 is a good way of correcting problems with Rmc. Sometimes the mud cake in front of the formation contains significant amounts of formation cuttings or lost circulation material and the F calculated is wrong. If the Micronormal is more than two times the Microinverse, use the empirical Figure 6-7 to

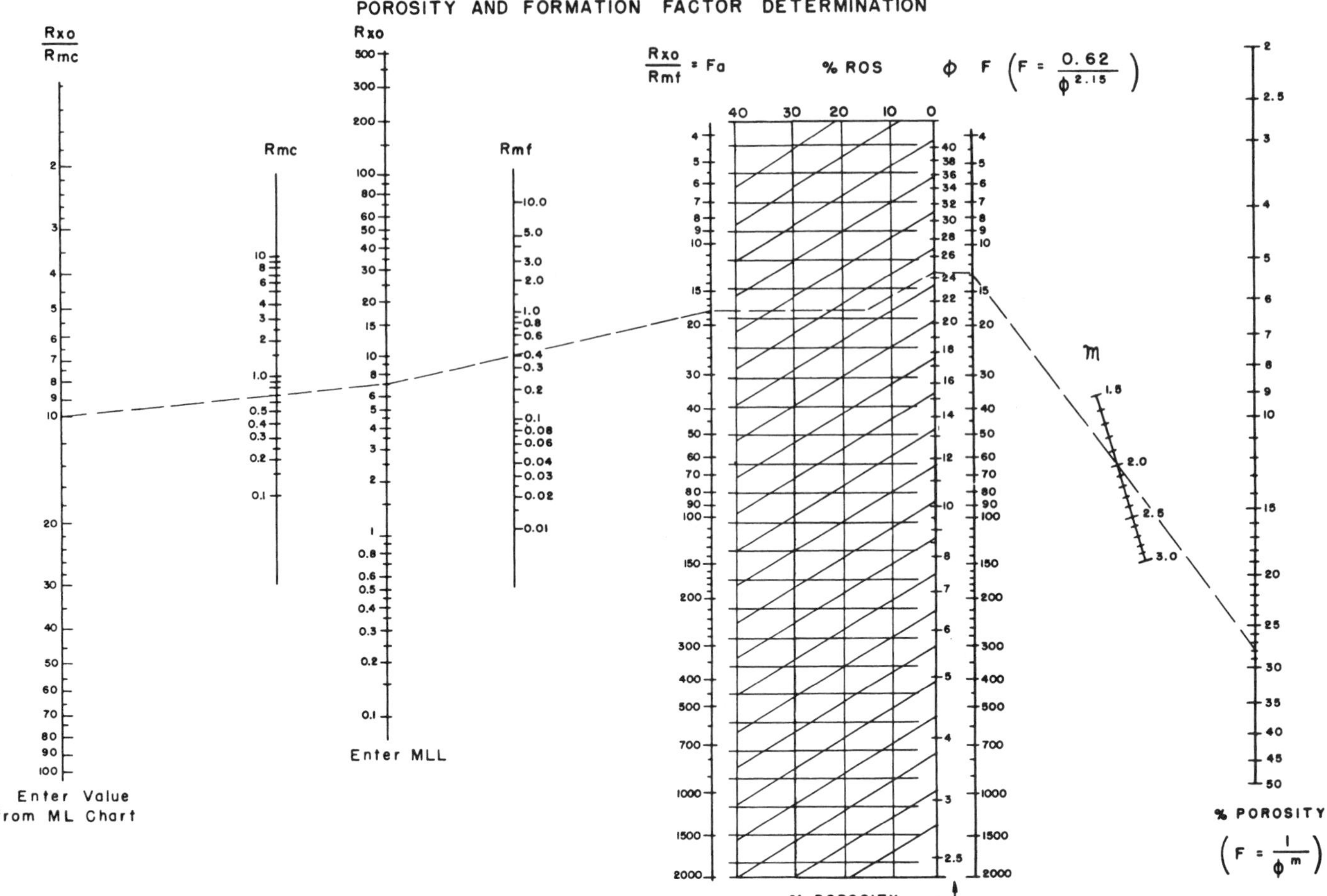

Figure 6-6.

obtain porosity. This usually results from current leakage and is considered by the author to be a poor Micrology (when the Micronormal is greater than two times the Microinverse resistivity reading)

If you are using the Micrology for porosity in low porosity formations, I suggest you check the values against the short normal technique in Chapter 7 and use the lowest porosity calculated. The short normal technique does not work well in soft rocks where the invasion is shallow.

MICROLOG MUD LOG

Usually, when the Microlog was run on the way into the borehole, the tool was collapsed. The resistivity curves were recorded to produce a mud log. The theory is that with the tool collapsed, the short investigating resistivity curves would see only mud in hole washouts and caves. This was used as a check on the Rm obtained and reported on the log heading. When the short and long spacings on the mud log read the same, this was the mud log mud resistivity at the indicated depth. Figure 6-9 is a typical mud log.

Where the long spacing (the dashed curve) is reading higher, it is being influenced by the formations. On the above log I would have picked Rm as 1.21 ohm m at about 5525 and not where the logging engineer picked it.

EXAMPLE LOGS

Example 6-1

This log is from Oklahoma. The caliper was drawn in Track #1. There are no obvious mud cake indications on the caliper. The Microlog shows two permeable intervals, 8647 – 65 and 8701 – 11. The upper section from 8600 to 8630 shows a severe borehole enlargement. The Microlog curves are reading just mud. The values agree with Rm on the heading.

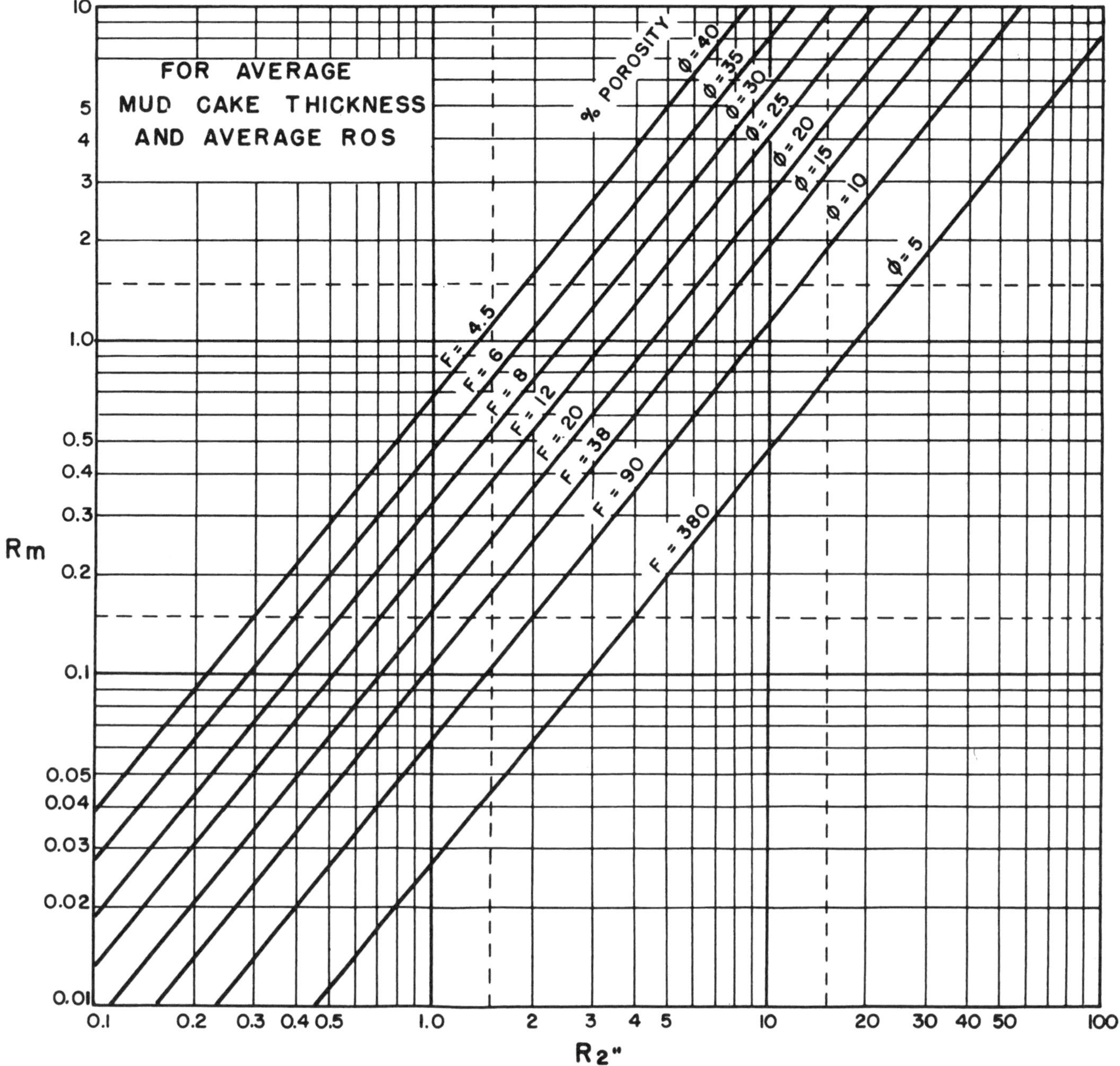

Figure 6-7. Empirical Microlog Chart (courtesy Schlumberger)

Porosity calculation 8654 – 8660

Rm = .52 at 135°F.
Rmf = .35 at 135°F.
Rmc = 1.00 from Figure 6-5

$R_{1\times1} = 2$ ohm m $\qquad R_2 = 3.5$ ohm m

$\frac{R_{1\times1}}{Rmc} = 2 \qquad \frac{R_2}{Rmc} = 3.5$

Entering Figure 6-4 with the above values Rxo/Rmc = 30 and tmc = 5/8″

Rxo = Rxo/Rmc × Rmc = 30 × 1 = 30 ohm m
Going through Figure 6-6 you will get Rxo = 30, F = 86 and for ROS = 0 Ø = 10%
and for ROS = 15 % Ø = 12%
Empirical Figure 6-7 gives you Ø = 15%

Example Log 6-2

Porosity calculation 7270 – 7290

Rm = .91 at 163°F
Rmf = .68 from Figure 6-5

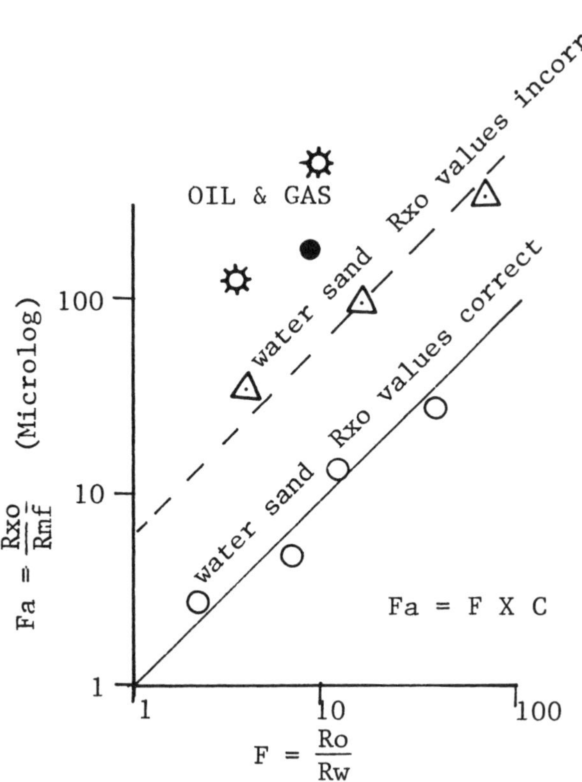

Figure 6-8. Microlog Fa Check

Rmc = .98 from Figure 6-5

$$\frac{R_{1\times1}}{Rmc} = \frac{1.4}{.98} = 1.43 \qquad \frac{R_2}{Rmc} = \frac{2.7}{.98} = 2.76$$

Rxo/Rmc = 100 from Figure 6-4 tmc = 1 inch
From Figure 6-6 F = 144 and for
ROS = 0 Ø = 8%
ROS = 15% Ø = 9.5%
These porosity values are too high. The porosity of these formations are in the neighborhood of 35%. Mud cake is probably too thick for the microinverse to work properly.

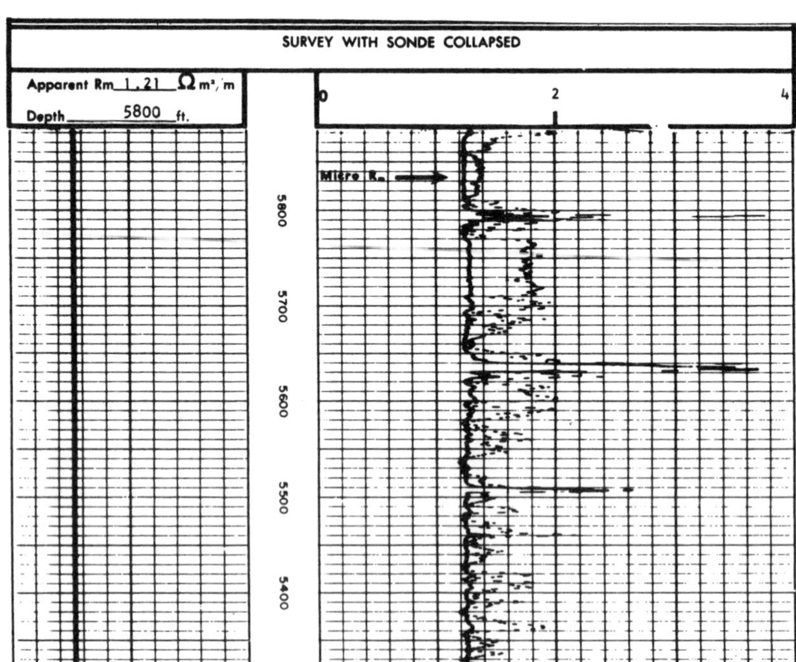

Figure 6-9. Microlog Mud Log

SPONTANEOUS POTENTIAL
20 MV
MICROCALIPER
6 7 8 9 10 11 12 13 14 15 16
Hole Diameter in Inches
SHORT NORMAL AM-16"
0 50
0 500
LATERAL AO-19'
0 50
0 500
LONG NORMAL AM-64"
0 50
0 500
Rm = .52 @ 135
Rmf = .35 @ 135
Rmc = 1.00 @ 135
0 Micro Inverse 1" x 1" 10
0 Micro Normal 2" 10
PERMEABILITY
8700

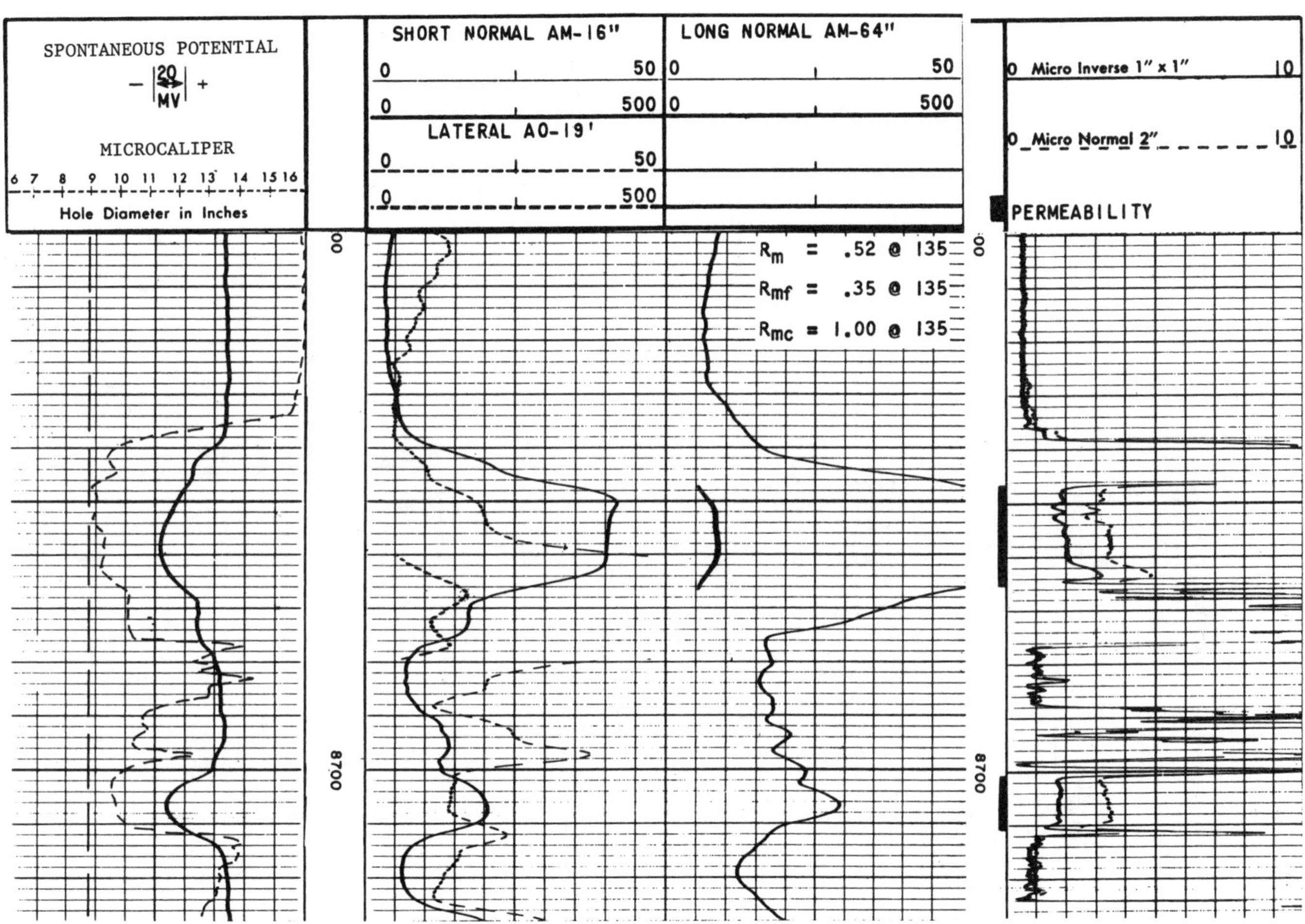

Example 6-1.

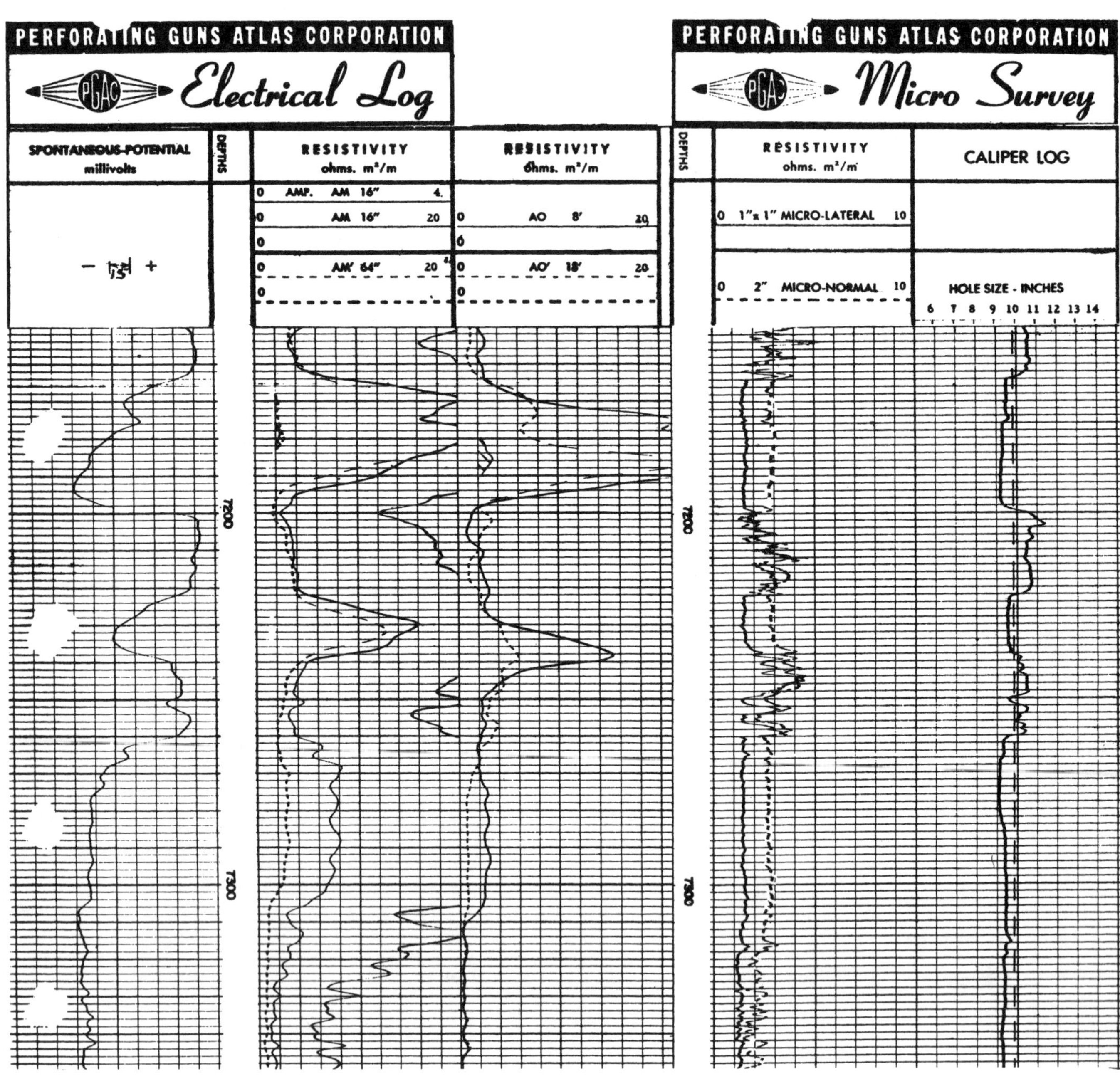

Example 6-2. ES and Microlog from DeWitt Co. Texas
Rm = 0.91 ohm m @ 163°F.

7

Hilchie, D. W., Porosity from the short normal, Old electrical log interpretation, p. 71–74.

Porosity from the Short Normal

In the Rocky Mountain technique discussed in Chapter 1, the ability of the short normal to measure porosity is implied. In the formations where the Microlog does not often give good porosities, another system was needed.

An estimate of porosity is possible from the short normal where invasion is normal or deep, as is common in most hard rock country, and the porosity is less than 25%. In other areas the short normal sees too much Rt due to shallow invasion and the empirical relationships do not hold.

THEORY

For a formation completely invaded by mud filtrate

$$S_{xo}^2 = \frac{F\ Rmf}{Rxo}$$

and if

$$F = \frac{1}{\emptyset^2}$$

then

$$Sxo^2 = \frac{Rmf}{\emptyset^2\ Rxo}$$

and

$$\emptyset = \frac{1}{Sxo}\sqrt{\frac{Rmf}{Rxo}} \qquad (7\text{-}1)$$

and since the short normal reads too deep to measure Rxo, this can be changed to

$$\emptyset = \frac{1}{Si}\sqrt{\frac{Rm}{Ri}} \qquad (7\text{-}2)$$

which is then converted to

$$\emptyset = C_{SSP}\sqrt{\frac{Rm}{Ri}} \qquad (7\text{-}3)$$

where C_{SSP} is an empirical constant that takes into account that the Rm should really be Rmf with a correction for formation water not flushed out of the field of vision of the short normal and also corrects for the oil saturation in the invaded zone when present. Figure 6-1 is a plot of C_{SSP} versus SP for water and oil bearing zones.

Figure 6-2 is a solution of equation 7-3 including the correction in Figure 6-1. If invasion is normal, Figure 6-2 may be used with few problems. If invasion is very deep, I use the water correction only. If invasion is too shallow, the porosities calculated will be too low.

EXAMPLE PROBLEM

Example log 7-1 is from the Denver Basin and was a by-passed oil well. The formation is the Muddy "J". The interval to be analyzed is 6654 – 69.

Data Taken From Log

$R_{SN} = 50$ $R_{LN} = 32$ R_L is not usable as this zone is in the dead zone caused by the resistive bed at 6645

SP = −80 mv.

e = 15 feet $R_{1\times1} = 3.2$ $R_2 = 6.1$

Rm = 1.5 @ 145°F

Calculated Data

Rmf = 1.23 Rmc = 1.35 from Figure 6-5

$$Rt = \frac{R_{LN} \times R_{LN}}{R_{SN}} = \frac{32 \times 32}{50} = 20.5$$

$$\frac{R_{1\times L}}{Rmc} = \frac{3.2}{1.35} = 2.37 \qquad \frac{R_2}{Rmc} = \frac{6.1}{1.35} = 4.52$$

from Figure 6-4 Rxo/Rmc = ∞ tmc = $^1/_2$ inch this will give 0% porosity

$$\frac{R_{SN}}{Rm} = \frac{50}{1.5} = 33$$ from Figure 3-8 Ri/Rm = 33

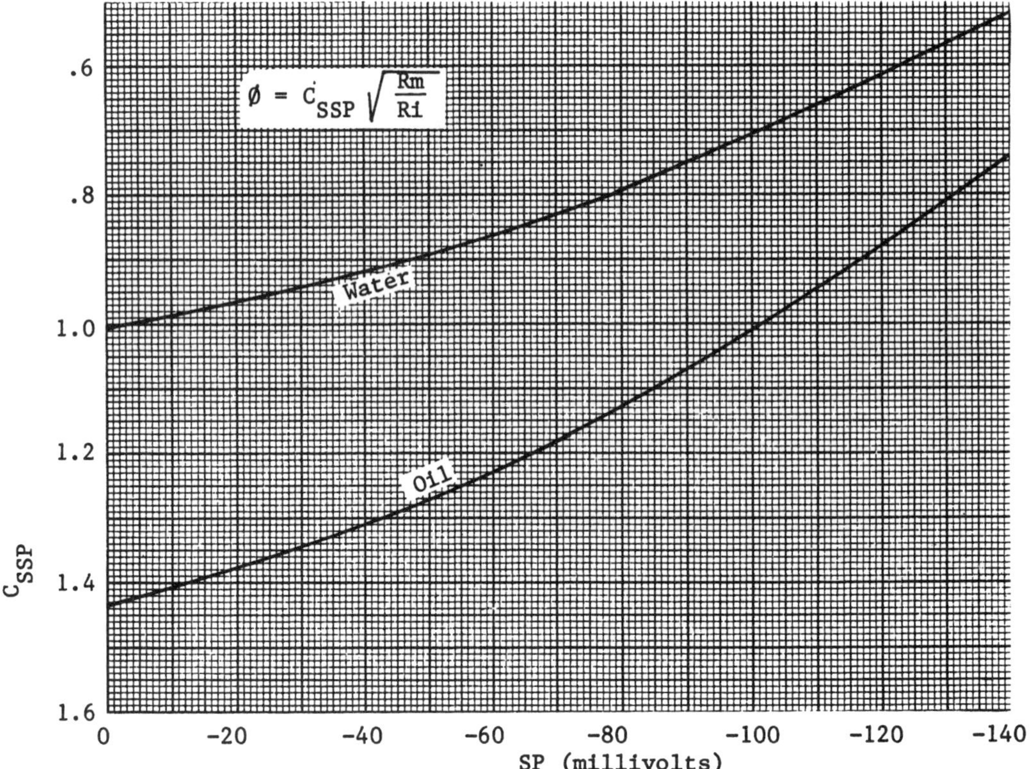

Figure 7-1. C_{SSP} versus SP for water and oil bearing formations

From Figure 7-2 for oil Ø = 17.5% for water Ø = 14%

Sw from Rocky Mountain Method in Chapter 1

Rmf/Rw from SP = 9.9 Ri/Rt = 50/20 = 2.5

Sw = 43% for normal invasion

Comparative Data from a new well 100 feet away where an induction and density log were run.

Porosity an almost constant 13%
Rt from induction log 24 ohm m
Rw = .1 from SP and Rwa technique
Sw = 46%

$$\phi = C_{SSP}\sqrt{R_m/R_i}$$

Note: Use for invaded non-shaly beds where $R_i/R_m \geq 20$ only.

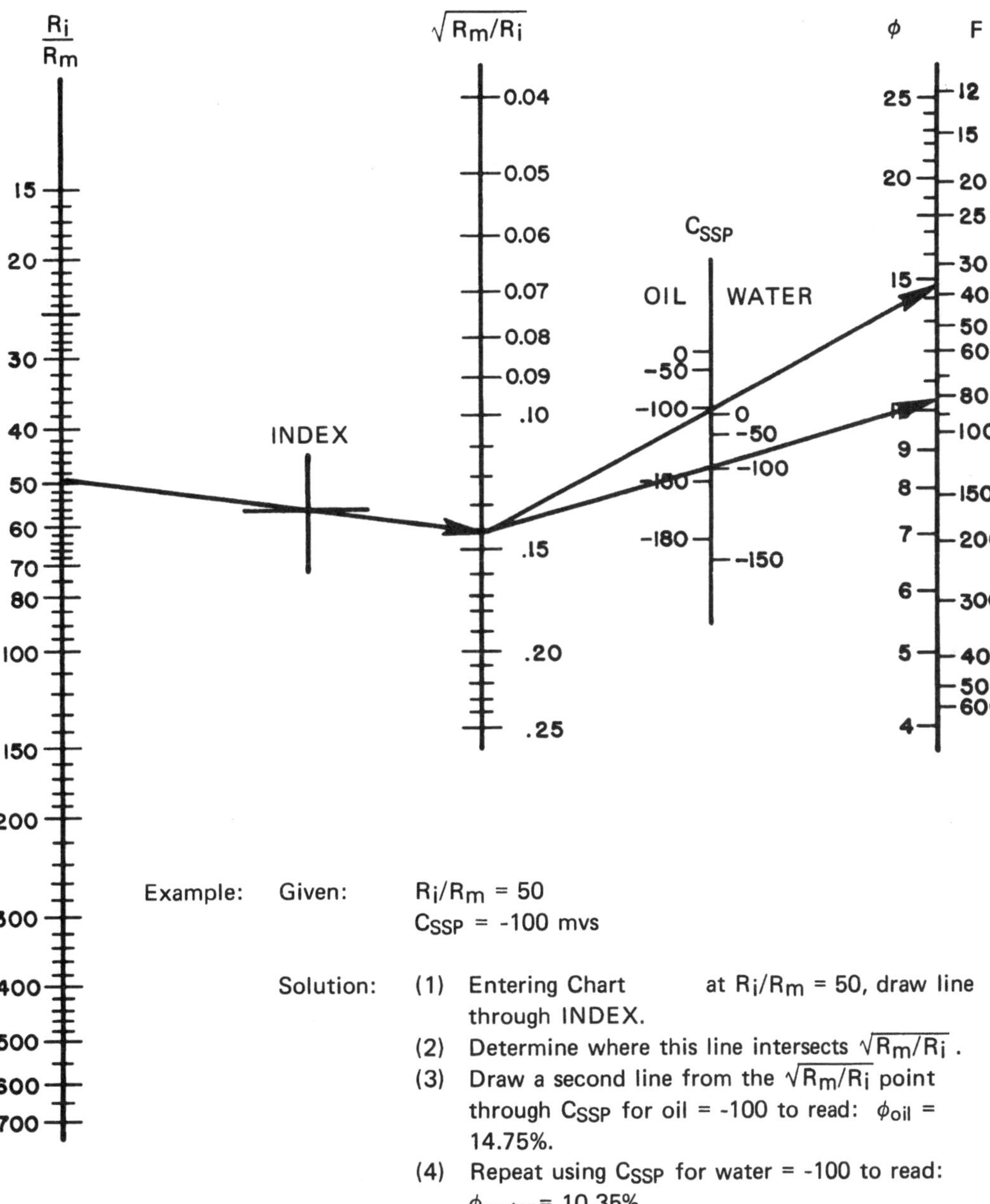

Example: Given: $R_i/R_m = 50$
C_{SSP} = -100 mvs

Solution:
(1) Entering Chart at $R_i/R_m = 50$, draw line through INDEX.
(2) Determine where this line intersects $\sqrt{R_m/R_i}$.
(3) Draw a second line from the $\sqrt{R_m/R_i}$ point through C_{SSP} for oil = -100 to read: ϕ_{oil} = 14.75%.
(4) Repeat using C_{SSP} for water = -100 to read: ϕ_{water} = 10.35%

Figure 7-2. Porosity from short normal nomograph (courtesy Birdwell)

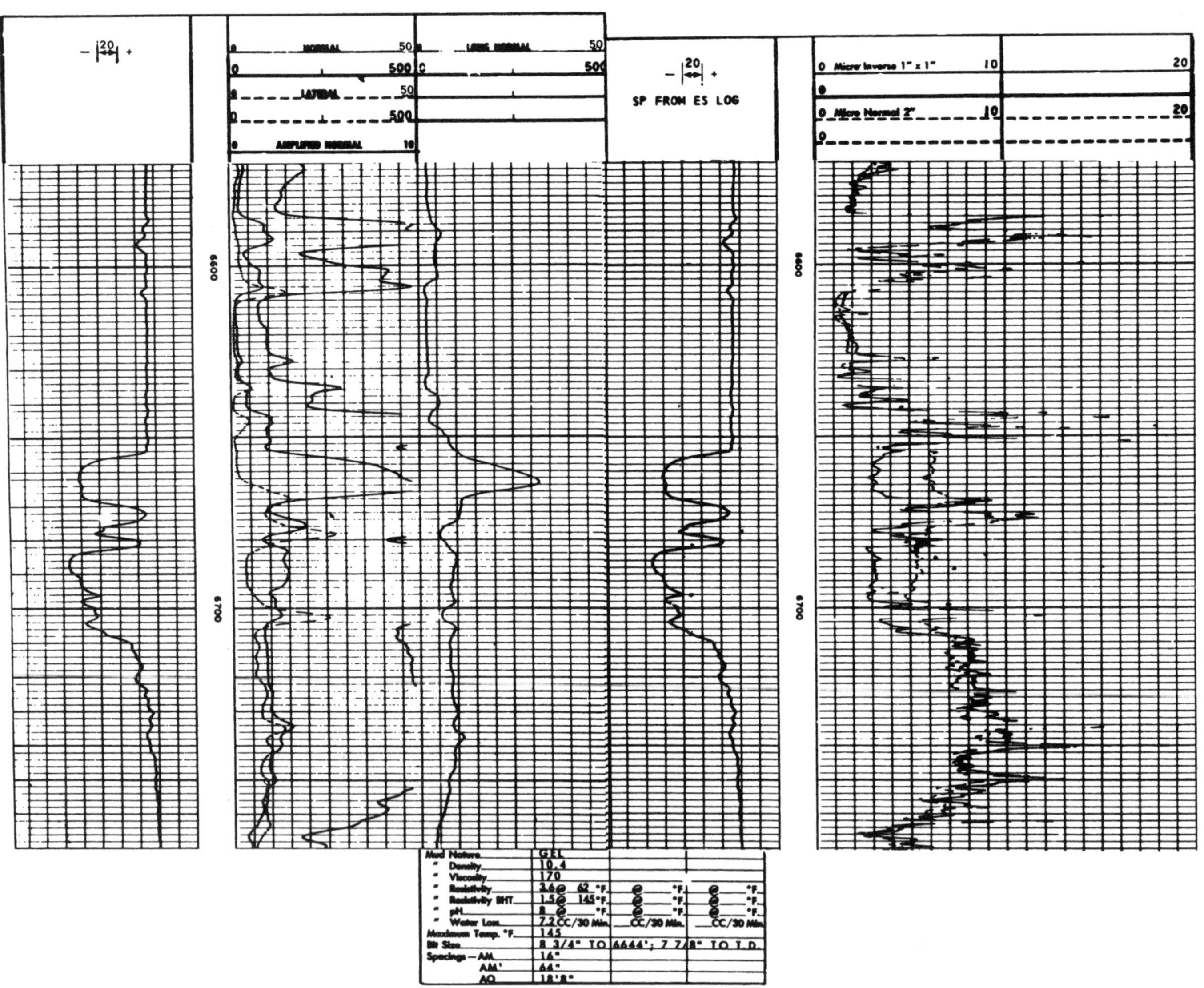

Example log 7-1

8

Hilchie, D. W., The limestone device, Old electrical log interpretation, p. 75–78.

The Limestone Device

The limestone device was developed in 1945 to overcome the distortions of the lateral curve. Its use was limited to West Texas and, to a much lesser degree, Alberta. The limestone device was used to determine porosity along the same lines as the short normal. It detected Ri and this was converted to porosity. The philosophy was that a short lateral measures Ri better than a short normal. The life and use of the limestone device was short, and as said before, geographically restricted. Few limestone devices were run after the early 1950's. The limestone device was run with a 10 inch normal and a 19 foot lateral. It was recorded in track #2 as a dashed curve behind the short normal.

The limestone device, as shown in Figure 8-1 is truly two laterals, one regular and one inverted, run simultaneously. The A0 spacing was shorted to 32″ (with some $37^1/_2$″ ones that were run with laterolog equipment). In reality, the current and potential electrodes (A and M respectively) were interchanged from Figure 8-1 as is allowed by the theory of reciprocity discussed in the section on laterals. It is easier to understand the operation of the tool using the system shown in Figure 8-1. The current is output through electrode A and returned to B. The voltage drop is measured between M and N. The curve presented is symmetrical.

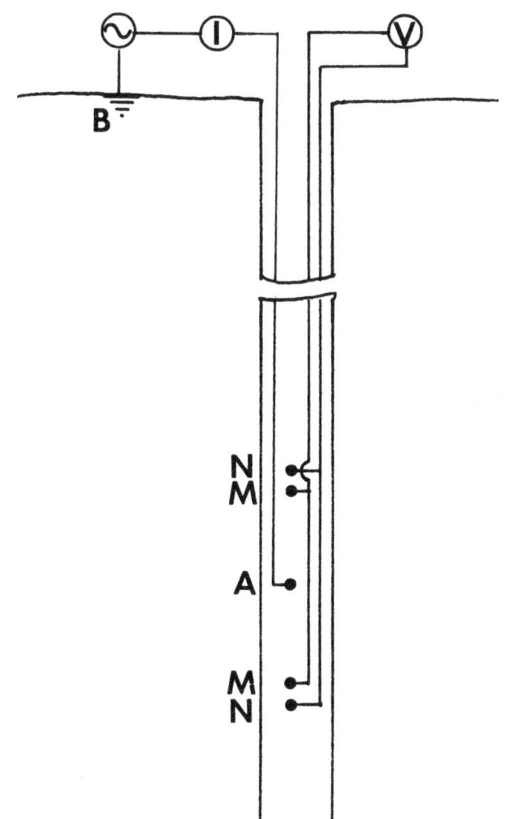

Figure 8-1. The Limestone Device

As with the normal, there is a maximum resistivity that the limestone device can measure. It is controlled by the mud resistivity (Rm) and the hole diameter, and, of course, the electrode spacings. Opposite a bed of infinite resistivity, the current from A flows up and down the borehole and the voltage drop seen by M and N is due just to the resistivity of the mud and the borehole diameter. The maximum resistivity that can be obtained is

$$\frac{\text{Ra (max)}}{\text{Rm}} = \frac{8\ \text{AM AN}}{d^2 - d_s^2}$$

where d is the borehole diameter, ds is the sonde diameter and AM is 30 inches and AN is 34 inches for the 32 inch limestone device. For a 3 5/8 inch diameter sonde used by Schlumberger, the following are for a 32 inch limestone device:

hole diameter	***Ra(max)/Rm***
4 3/4″	866
6 3/4″	252
8″	160
9″	120
10″	94
12″	62

This means that in a 9 inch borehole filled with .5 ohm m mud

$$\frac{\text{Ra(max)}}{\text{Rm}} = 120$$

$$\begin{aligned} \text{Ra(max)} &= \text{Ra(max)/Rm} \times \text{Rm} \\ &= 120 \times .5 \\ &= 60 \text{ ohm m.} \end{aligned}$$

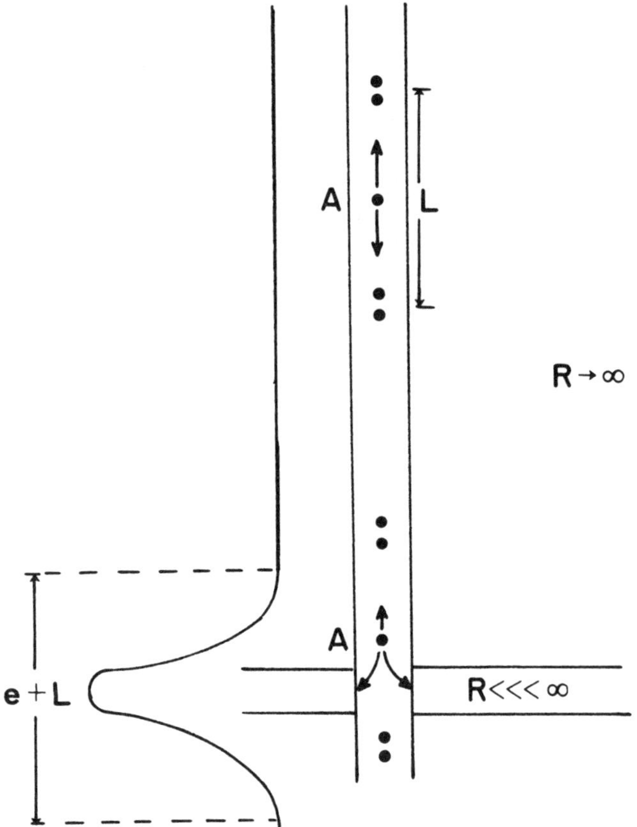

Figure 8-2. The Limestone Device Response to a Conductive Bed

If the borehole diameter and mud data are right, the maximum reading on the limestone device in a zone of essentially infinite resistivity will be 60 ohm m.

Figure 8-2 shows the limestone device response to a bed with infinite resistivity and a bed with a much lower resistivity. In the upper part of the example the current flow is equally distributed up and down the borehole and the potential drop is controlled by the Rm and hole diameter. Opposite the conductive bed, the current flows along the lowest resistance path (into the conductive bed) and the voltage drop between M and N is reduced as the current flow is reduced. The conductive bed shows on the log to have an anomaly equal to the bed thickness plus the length of the electrode array.

POROSITY FROM THE LIMESTONE DEVICE

In reality Ri/Rm is obtained from the limestone device and this is converted to porosity in much the same manner as was done for the short normal.

Two methods may be used to obtain porosity from the limestone curve. The first is the Relative Deflection method proposed by Schlumberger and the other is the crossplotting of the 10 inch normal and the limestone as proposed by S. J. Pirson.

Porosity using the Relative Deflection Method

Figure 8-3 is the chart used for relative deflection method. To use this chart, you determine the Ra (max) in a tite zone on the log or calculate the Ra (max) from the previous section (Table 8-1). The deflection in the zone of interest is obtained from the log. The percent deflection is calculated. You enter the chart with borehole diameter (bit size if there is no caliper) and percent deflection and come out with Ri/Rm and porosity. Correct the porosity for hydrocarbons, etc., using C_{SSP} system discussed in Chapter 7, Figure 7-2.

Porosity using the Crossplot Method

Figure 8-4 is a crossplot of the 32 inch limestone curve versus the 10 inch normal. The combination of both devices gives you both Ri/Rm and borehole diameter. If the mud resistivity or borehole diameter is wrong, the point obtained by entering the chart with $R_{10''}$/Rm and $R_{32''}$/Rm will indicate an erroneous borehole diameter (d). To correct for either of these problems, move the point plotted at 45° to the correct (d) and read Ri/Rm. With Ri/Rm go to Figure 7-2 and solve for porosity just as you did for the short normal. I have had some problems with this approach. I believe it is because both devices are reading the same amount of formation and thus do not compliment each other.

Example 8-1

Example Log 8-1 is a West Texas Limestone Device and Microlog.

The maximum reading on the limestone (1s) is 48 ohm m. Table 8-1 gives a maximum value (Ra (max)/Rm × Rm) of 60 ohm m for the 9 inch borehole.

Maximum Deflection Method

interval 8386
Ra from limestone device is 16 ohm m
using apparent maximum of 48
% deflection = 16/48 = .33
on Figure 8-3 this is 19% porosity
using Table 8-1 maximum of 60
% deflection = 16/60 = .27
on Figure 8-3 this is 22% porosity
empirical Microlog chart Figure 6-7 gives ($R_{2''}$ = 3 and Rm = .5) 20% porosity

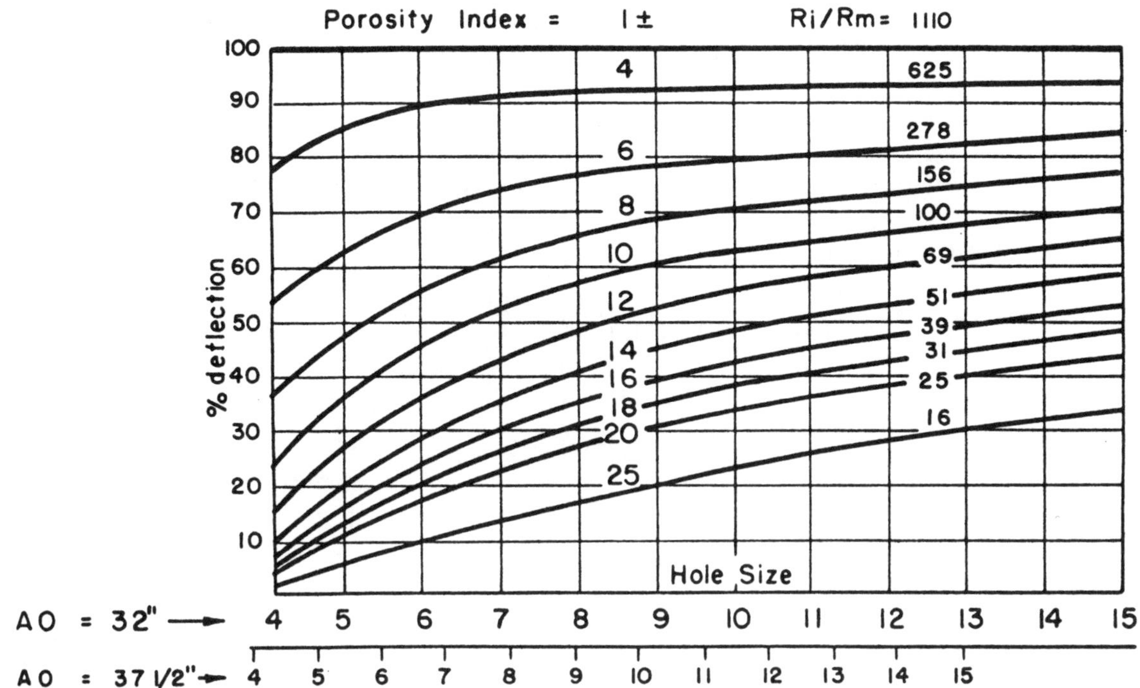

Figure 8-3. Relative Deflection Method

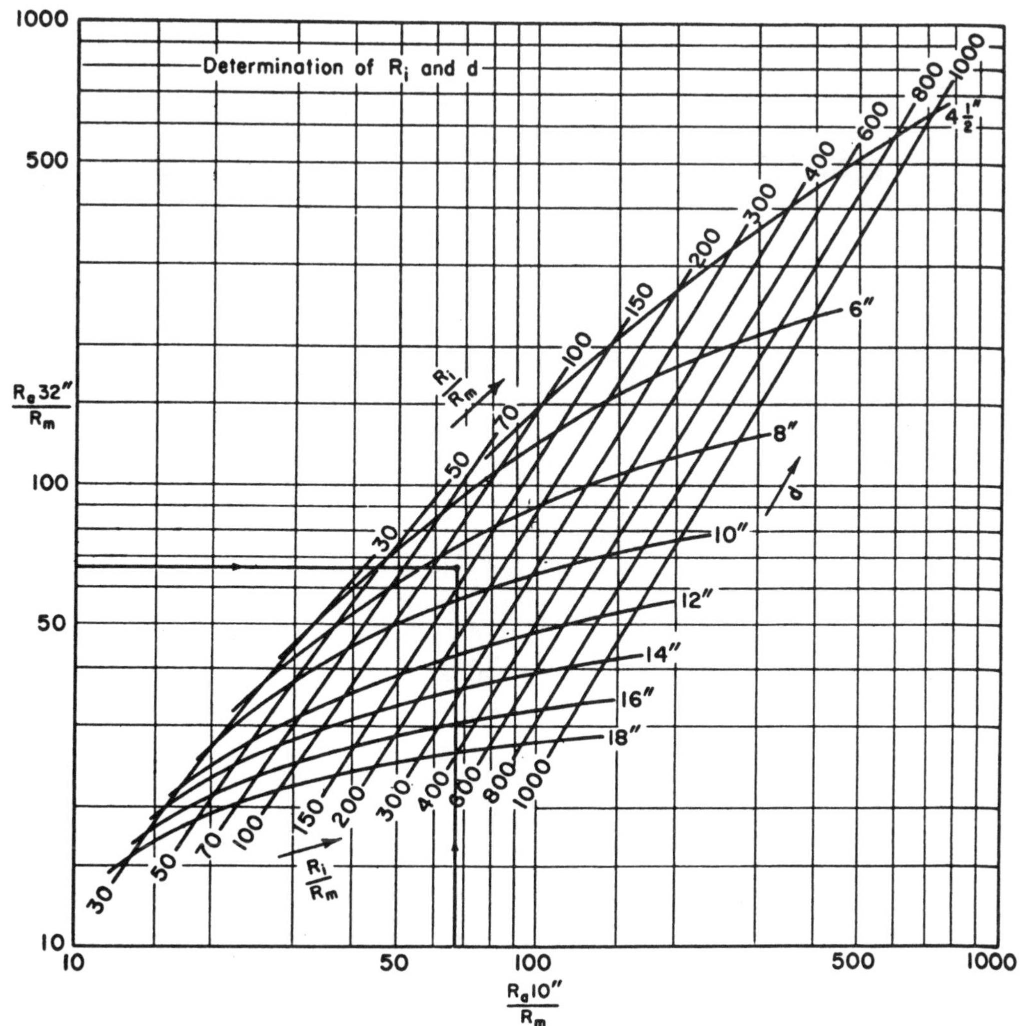

Figure 8-4. Limestone and 10 Inch Short Normal Crossplot (from S. J. Pirson World Oil, April 1957)

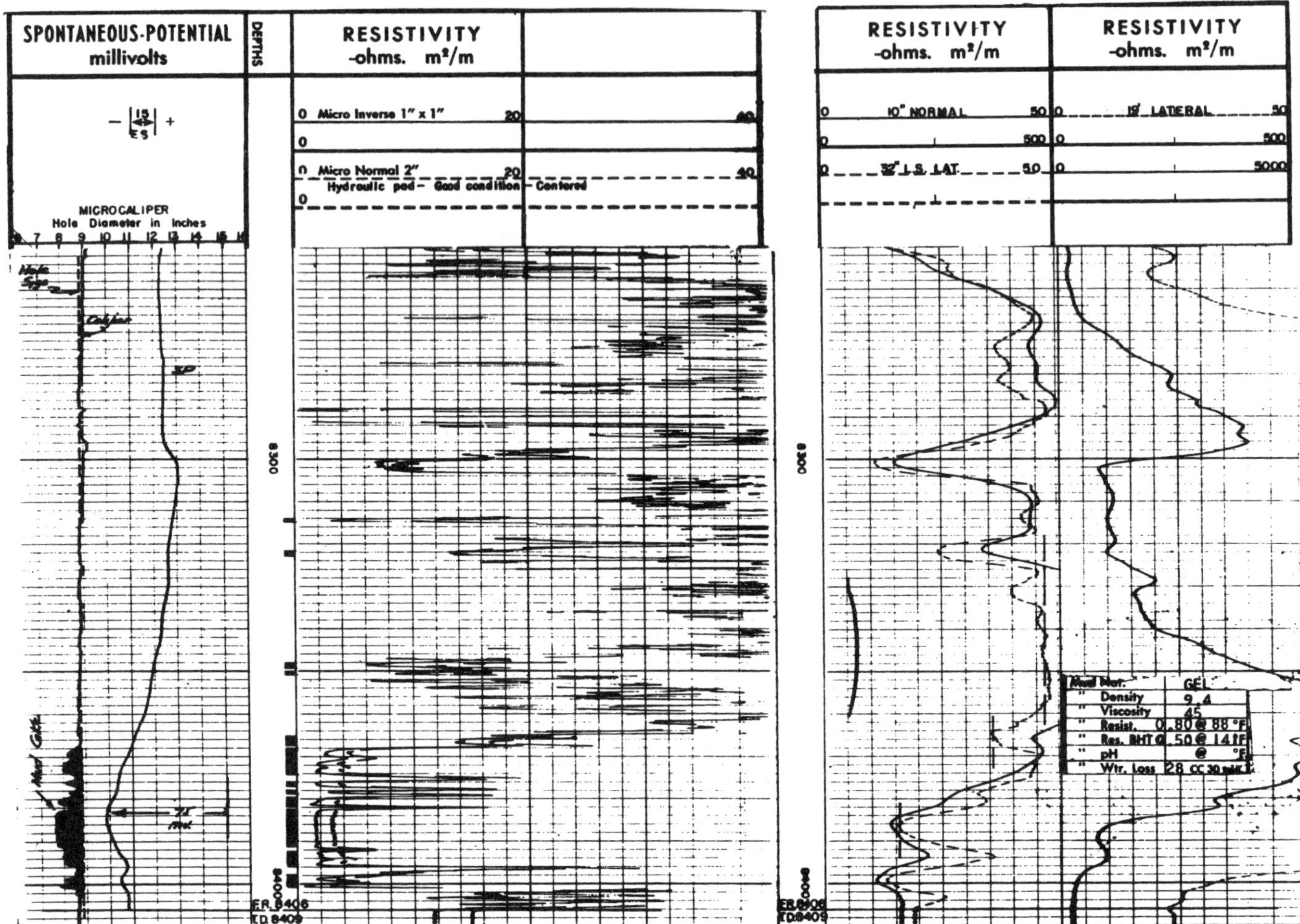

Example Log 8-1.

Crossplot Method

interval 8386
Ra from limestone device is 16 ohm m
  Ra/Rm = 16/.5 = 32
Ra from short normal is 14 ohm m.
  Ra/Rm = 14/.5 = 28
from crossplot Ri/Rm = 90 which from Figure 7-2 porosity is 11%

There is severe disagreement between the two methods in this example. The short normal by itself using Chapter 7 approach gives a porosity range of 15 to 22%, which tends to make the relative deflection method more believable for this example.

COMMENT

The limestone device was not successful in West Texas as a porosity device and most people I have associated with over the years do not believe the porosities calculated from it. I believe the limestone device is like the Micrölog, you only use it for porosity when you are desperate.

9

Hilchie, D. W., Salt mud surveys — Laterologs and microlaterologs, Old electrical log interpretation, p. 79–83.

Salt Mud Surveys — Laterologs and Microlaterologs

In salt mud filled boreholes the ES did not work too well and so salt mud equivalents to the ES and Microlog were introduced in about 1950.

THE LATEROLOG

The laterologs (called a Guard log by Welex and also intoduced in about 1950) used in the 1950's were either 7 or 3 electrode devices usually called Laterolog 7's (LL7) or Laterolog 3's (LL3). Figure 9-1 shows a schematic of both electrode arrays.

The principle behind laterologs is to force a beam of current into the formation by use of guard electrodes emitting the same polarity current. On the LL3 the two large guard electrodes on each side of the survey current electrode did this. On the LL7 two monitor electrodes between the guard and survey electrode controlled the output of the guard electrodes. The potential between the two monitoring electrodes was kept at zero by increasing and decreasing the output from the guard electrodes. If the potential between the monitoring electrodes was zero, no current was going up or down the hole and all the surveying current was going into the formation. The current beam widths ranged from 3 inches to almost three feet. This gave the laterologs very good thin bed resolution. The vertical resolution is slightly larger than the surveying current beam width. The laterologs could thus "see" beds down to one and two feet thick.

Only one resistivity curve was run and this was often assumed to be Rt and used accordingly. In reality laterologs are significantly influenced by invasion. For example, the new dual laterologs, which have depths of investigation of twice the old laterologs, require corrections of 30–50% in modern well log interpretation. These indicate an average invasion diameter of about 40 inches. In the older laterologs this represents about a correction of 70% because of the shallower depth of investigation. If this 40 inch invasion diameter is assumed to be a constant, then these old laterologs should have their apparent resistivity (Ra) from the curve increased as:

$$Rt = 1.67\ Ra - .67\ Rxo \qquad (9\text{-}1)$$

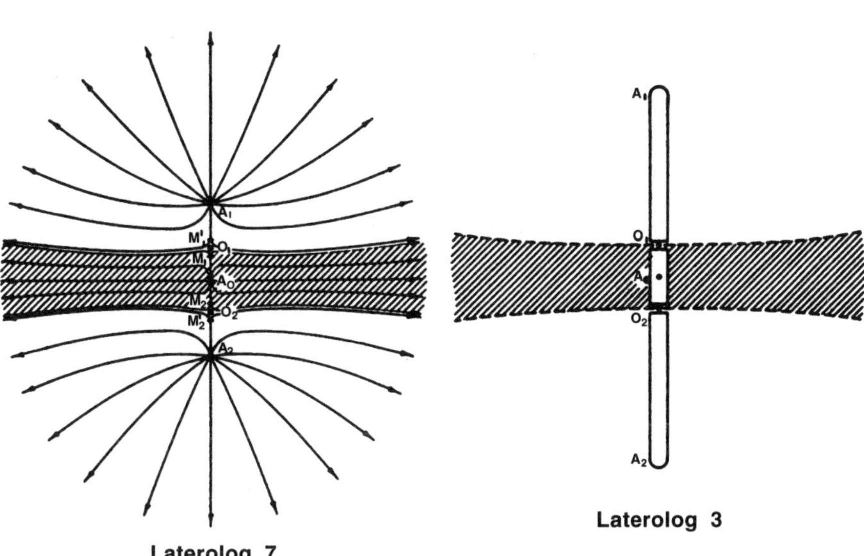

Figure 9-1. Electrode Schematics of Laterolog 7 and Laterolog 3 Instruments

where Ra is the value from the laterolog curve and Rxo is obtained from the microlaterolog to be discussed later in this section. The correction shown in equation 9-1 is significant. If you do not use it in hydrocarbon bearing zones, the calculated water saturations will be too high and the cut-off water saturations must be adjusted upward as they were in many older log interpretations. Table 9-1 shows an example of the influence of the low resistivity readings from uncorrected old laterologs.

Table 9-1 shows that at high water saturations, there is little difference; but as the water saturation decreases,

Table 9-1. Calculations showing the effects of Sw of low Rt estimations

Ra (from laterolog) = 3 ohm m	Sw = 48%
Rt from eqn. 9-1 = 4.3	Sw = 40%
Ra (from laterolog) = 1.3	Sw = 73%
Rt from eqn. 9-1 = 1.5	Sw = 68%

Assumed data — Rxo = 1 ohm m Rmf = Rw Ro = 0.7 ohm m

the error increases. The error is significantly increased if Rmf is either larger or smaller than Rw.

Note: laterologs run in fresh muds where Rmf is significantly greater than Rw result in laterolog readings very close to Ri values.

THE MICROLATEROLOG

The microlaterolog is a pad version of the LL7. A cylindrical beam of current is forced into the formation by circular guard electrodes as shown in Figure 9-2. The current only penetrates about 2 to 4 inches and reads the Rxo or flushed zone. The flushed zone is the part of the formation close to the borehole wall in which all the formation water has been flushed out. Some of the movable hydrocarbons are also flushed out. The microlaterolog is a salt mud tool as it requires thin mud cakes to read properly. It is usually assumed to read Rxo directly with no corrections necessary. If the mudcake is too thick, the Rxo value will be too low. Unfortunately, it is very difficult to determine the mudcake thickness and, thus, also difficult to correct properly for mudcake influences. So we don't do it.

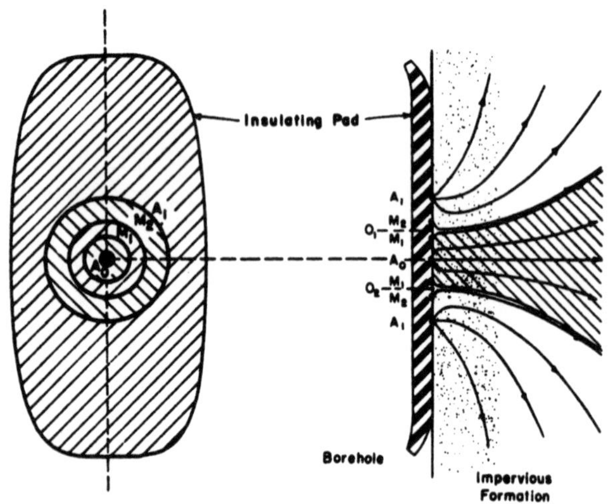

Figure 9-2. Schematic Microlaterolog Electrode Array

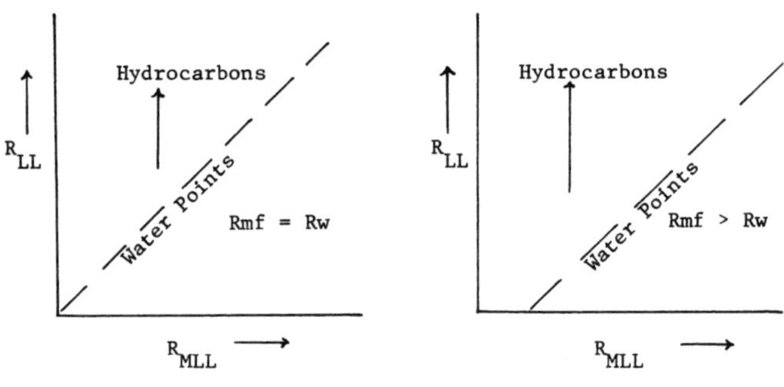

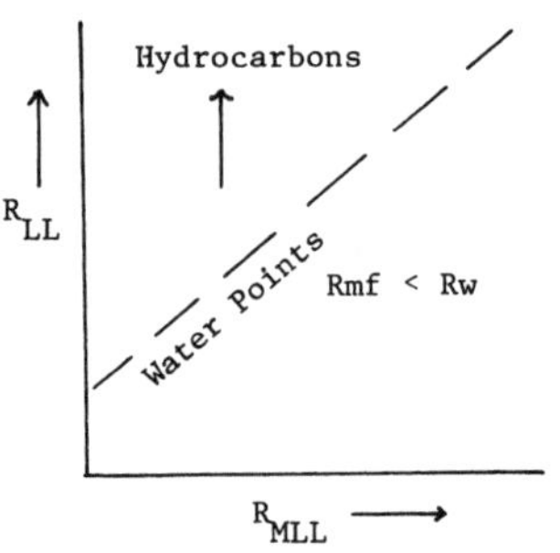

Figure 9-3. Laterolog - Microlaterolog Crossplots (LL - MLL)

The microlaterolog was often called a 'Trumpet log' because of the way the current beam spread resembled the end of a trumpet. Other trade names for this type of device are FoRxo, Mini-Focused and Micro-Laterolog.

As there were no porosity logs in the period of time being discussed, the microlaterolog and laterolog were used together in interpretation schemes.

LATEROLOG AND MICROLATEROLOG INTERPRETATION

Combined laterolog and microlaterolog interpretation works on the basis that the hydrocarbon saturation in the flushed zone that the microlaterolog sees is lower than in the virgin zone. Without this difference, the techniques will not work. Typically, the resistivity from the laterolog was plotted versus the microlaterolog resistivity as shown schematically in Figure 9-3. If Rw = Rmf, the laterolog and microlaterolog would be the same in water zones and the laterolog resistivity would be higher in the hydrocarbon bearing zones. If Rmf and Rw were different, the water points on the graph would still fall lower than the hydrocarbon zones. The difference is a function of the difference in hydrocarbon saturation in the flushed versus the virgin zones.

Mathematically, the response of the laterolog (assuming it reads Rt) is

$$Rt = \frac{F\ Rw}{Sw^2} \tag{9-2}$$

and the microlaterolog is:

$$Rxo = \frac{F\ Rmf}{Sxo^2} \tag{9-3}$$

where

Rxo is the resistivity of the flushed zone (from the microlaterolog)

Rmf is the mud filtrate resistivity

Sxo is the water saturation in the flushed zone.

Ratio equations 9-2 and 9-3 and eliminating the common F and reorganizing

$$\frac{Sw^2}{Sxo^2} = \frac{\dfrac{Rxo}{Rt}}{\dfrac{Rmf}{Rw}} \tag{9-4}$$

and if

$$Sxo^5 = Sw$$

as found by Schlumberger in empirical studies then:

$$Sw = \left(\frac{\dfrac{Rxo}{Rt}}{\dfrac{Rmf}{Rw}}\right)^{\frac{5}{8}} \tag{9-5}$$

Equation 9-5 or the nomograph based on this equation, Figure 9-4, can be used to solve for water saturation if Rw, Rmf, Rxo and Rt are known.

I would suggest that the correction (equation 9-1) be applied to all laterolog resistivities in hydrocarbon bearing zones before solving for water saturation.

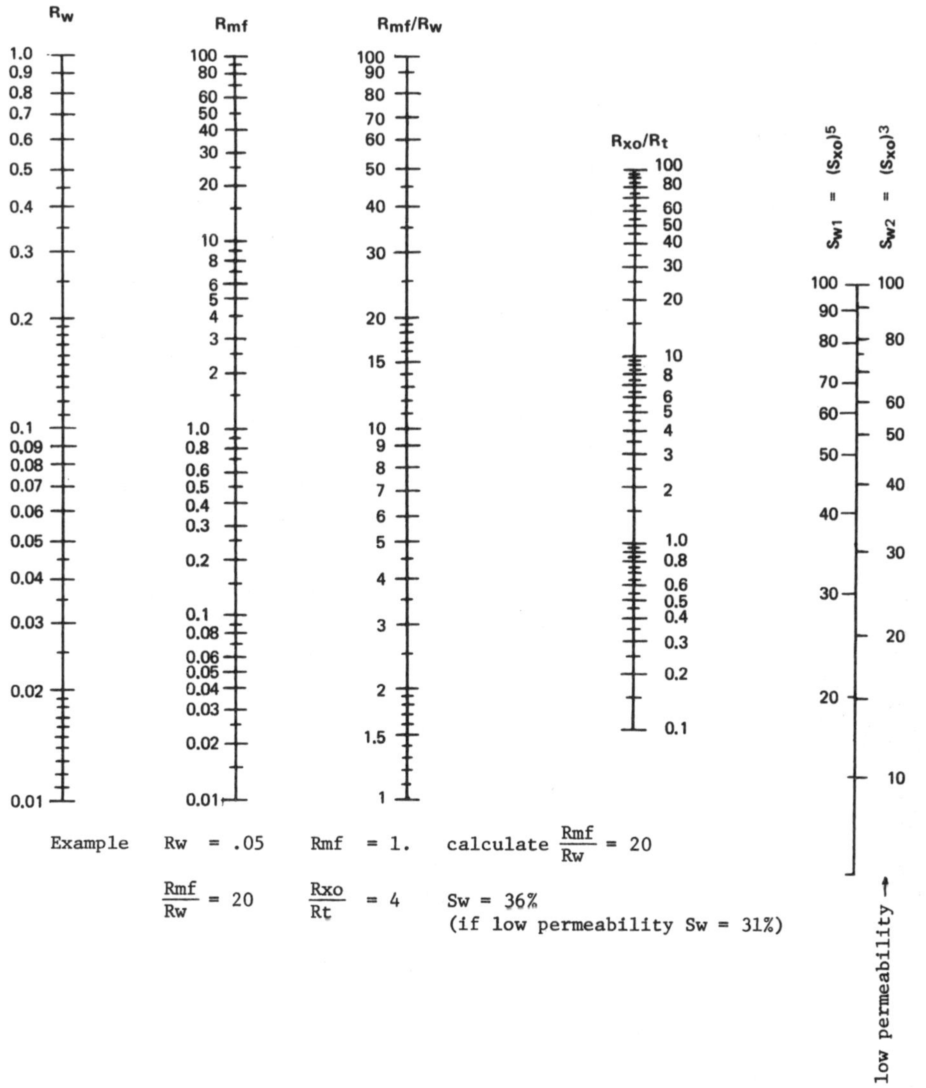
Rw
Rmf
Rmf/Rw
Rxo/Rt
Sw1 = (Sxo)5
Sw2 = (Sxo)3
Example Rw = .05 Rmf = 1. calculate Rmf/Rw = 20
Rmf/Rw = 20 Rxo/Rt = 4 Sw = 36%
(if low permeability Sw = 31%)
low permeability

Figure 9-4. Ratio Interpretation Scheme using equation 9-5 (courtesy Birdwell)

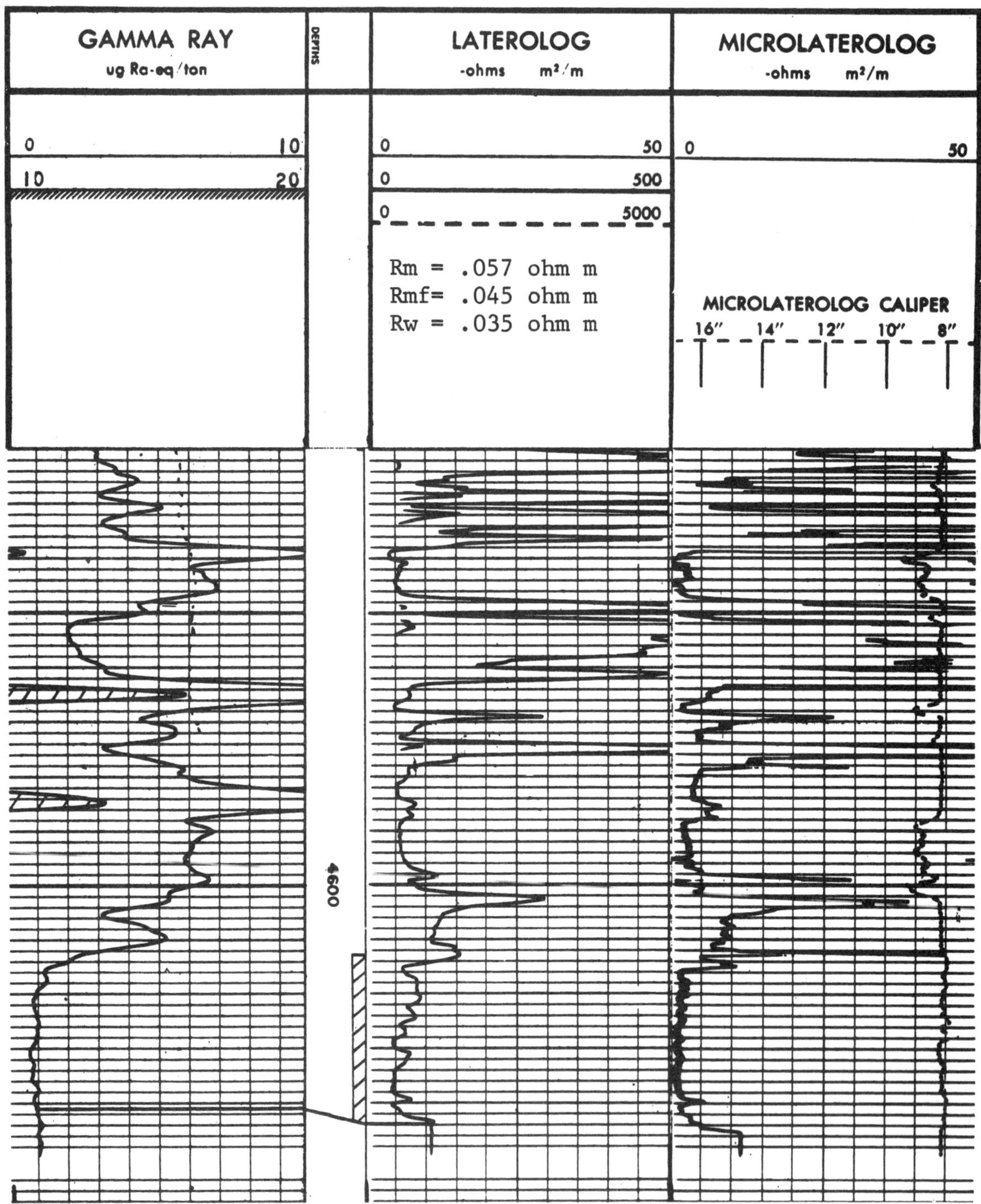

Example 9-1. Laterolog Example

Interval 4617–22

R_{LL} = 8.5 R_{MLL} = Rxo = 2

Using raw log data Rmf/Rw = 1.29   Rxo/Rt = .24   Sw = 34% (Fig. 9-4)

Correcting LL with eqn. 9-1   Rt = 12.9   Rxo/Rt = .16   Sw = 27%

Interval 4613–4643 I.P. 9.5 MMSCF/D No water.

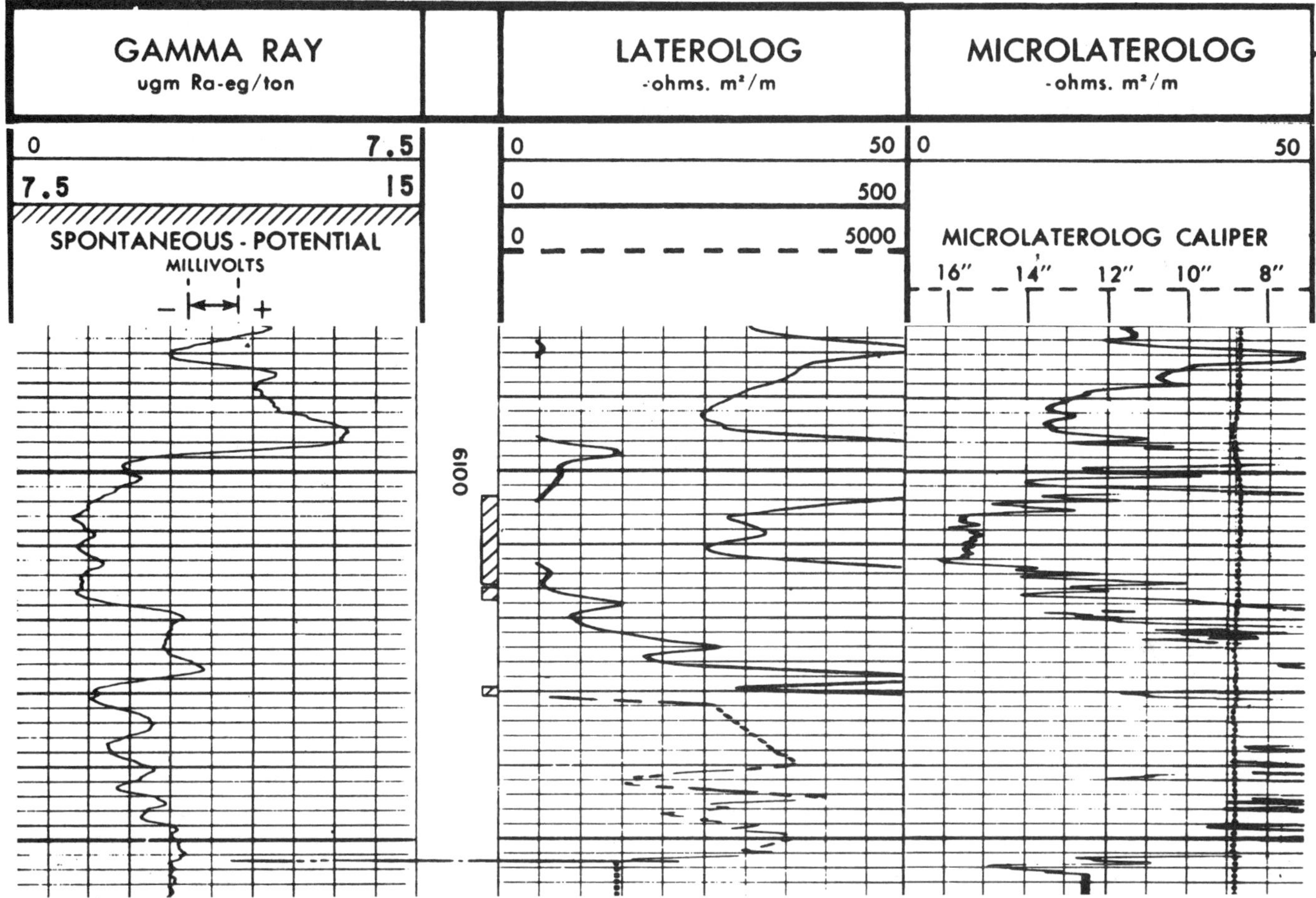

Example 9-2. Penn. Lime Oklahoma Example

Rm = .06 Rw = .04 Rmf = .05 (Figure 6-5)

Rmf/Rw = 1.25 Interval 6106 – 6112 R_{MLL} = Rxo = 8 R_{LL} = 28 (averaged)

Sw from Figure 9-4 (raw data) = 39% (6088 – 6120 produced 418 BOPD)

10

Hilchie, D. W., The old gamma ray and neutron logs, Old electrical log interpretation, p. 85–93.

The Old Gamma Ray and Neutron Logs

This chapter on old gamma ray and neutron logs is to expose you to the differences in these old devices as compared to the newer logs. Neutron logs came of age in the early 1960's just as did other logs. Previous to the 1960's, the logs were run too fast (reducing the vertical resolution) and most of the counters were not too efficient, resulting in relatively high statistical problems. In short, they were not really quantitative.

THE GAMMA RAY LOG

Since its inception in 1939 by Well Surveys Inc., the gamma ray has been used primarily as a qualitative shale log. The early logs used geiger counters or ionization chambers to detect the natural radioactivity of the formations. These counters were often long (up to 3 feet) and inefficient, resulting in poor thin bed resolution. They were often run very fast and resulted in curves that looked like the SP.

The gamma ray log responds primarily to shale, which is the most common high radioactivity formation we encounter. High count rates are shales while low count rates are nonshales (like sandstone, limestone, dolomite, anhydrite, etc). There are a few exceptions like granite (wash), locally radioactive sandstones and carbonates. Figure 10-1 shows a schematic gamma ray curve for most commonly encountered formations.

Gamma ray logs were calibrated in many different scales before the introduction of the API Unit which is tied to calibration pits in Houston at the University of Houston. Table 10-1 is an approximate conversion between the different units and API Units for older gamma ray logs.

THE NEUTRON LOG

The neutron log of the 1940's and 1950's was a cylindrical tube which at the lower end contained a neutron source (usually radium beryllium), on top of which was a shield and then a detector. The detector detected gamma rays of capture (neutron gamma), thermal neutrons (neutron neutron) or epithermal neutrons. The source to detector spacings were from about 10 to 20 inches, depending upon the tool and the service company.

The neutron log was a porosity log. It primarily detects hydrogen, which is directly relatable to porosity unless you are in a gas zone. Gas has low hydrogen content and shows up as low porosity. The neutron logs of this period were recorded in counts per second, environmental units, standard counts per second, etc. For most neutron logs in this time period the counting rate was approximately a linear relationship to the logarithm of porosity. Figure 10-2 is a calibration curve for one of these logs. Notice that the calibration curve is a straight line on a semilog plot of porosity versus counting rate. The sensitivity of the log to borehole size is obvious from Figure 10-2. Also notice that the different borehole sizes converge to essentially a common point at 40–45% porosity. The response of the measurement is very sensitive to porosity at low porosities (less than 10%) and insensitive at higher porosities (25–40%). This figure (10-2) is from "Radiation Logging" by J. Tittman, Petroleum Engineering Conference, University of Kansas, 1956, and is one of the best basic discussions of neutron logging ever published.

Neutron logs in this era gave relatively good porosity results in openhole applications but only fair results in cased holes. The problems were unknown lithology responses for the measurements, mudcake and borehole diameter correction problems and calibration problems. Figure 10-3 shows a conventional neutron log (with gamma ray) that has a porosity scale on the heading and shows core porosities for comparison. For an overall neutron responds refer back to Figure 10-1. Figure 10-4 shows a calibration curve developed for a particular formation using core analysis porosity.

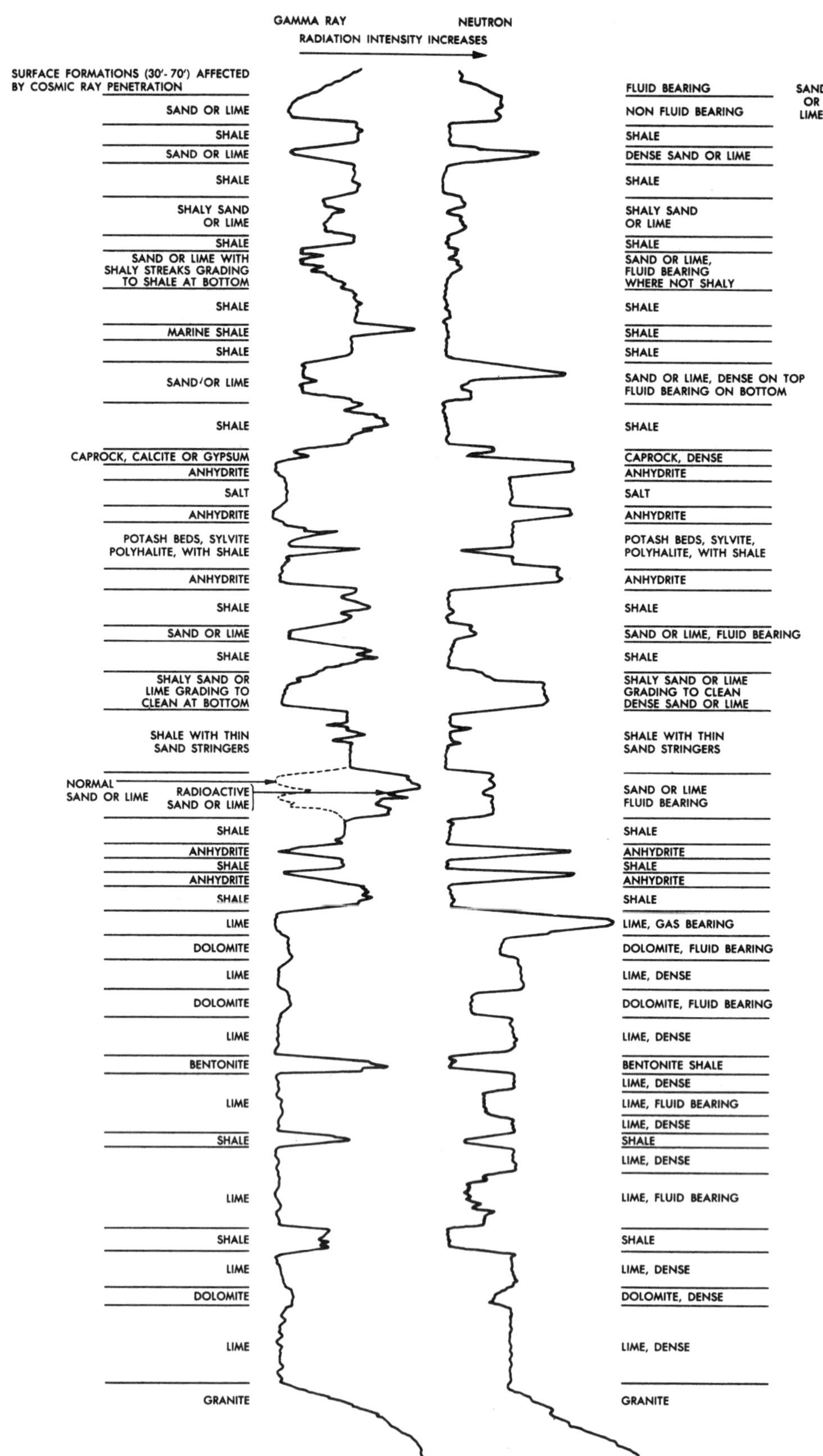

Figure 10-1. Gamma Ray and Neutron Responses (courtesy of Dresser Atlas)

Table 10-1. Old Gamma Ray Log Conversion Units

SERVICE COMPANY	*CONVENTIONAL UNITS*	*API UNITS*
Schlumberger	1 ug Ra equiv./ton	16.5
Lane Wells		
Series 400 (scintillation)	1 radiation unit	2.16
Series 300 (geiger counter)	20.2 counts /minute	1.
Series 200 (ionization ch.)	1 standard unit	216.
PGAC		
Type F (geiger counter)	1 microroentgen/hr.	14.
Type T (scintillation)	1 microroentgen/hr.	15
McCullough	1 microroentgen/hr.	10.4

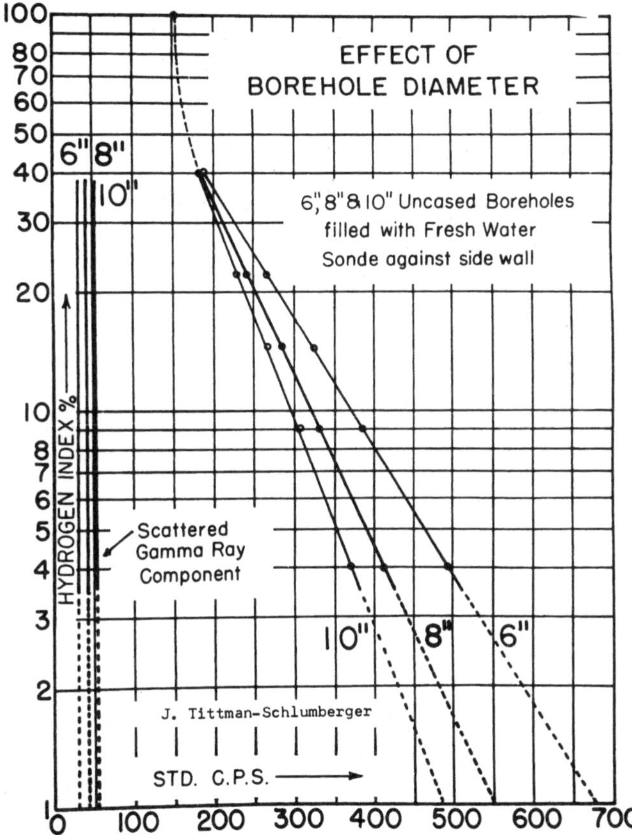

Figure 10-2. Calibration Curve for a Conventional Neutron Log

POROSITY FROM THE NEUTRON LOG

Calibration curves were available for some of the service companies for their neutron logs but most analysts found that relative calibration methods worked best most of the time. The principle was based on the fact that in apparent high porosities, the curves tended to saturate at 40 – 45% porosity. This means that if there was a borehole washout or in a shale the log counting rate could be assumed to be 40 to 45% apparent porosity. In reality, the analysts usually found the lowest counting rate on the log and used this counting rate as 45% porosity. A tite zone such as an anhydrite or lime stringer was used as a low porosity calibrator. These tite zones were (locally) 1, 2 or 3% porosity. You now have two points that you can plot on a semilog plot of porosity and count rate. Join these two points and you have the porosity calibration curve for the neutron log. If you do not have a tite zone, choose a low porosity water zone and calculate the porosity from the resistivity log and use this point.

Figure 10-5 is a blank neutron calibration graph.

Figure 10-6 is a Schlumberger Calibration graph on which the pivot points are already plotted. To use these pivot points (instead of the usual shale points),you must be sure the log is calibrated right. Note the different points in Figure 10-6 for the different sources. This is due to the difference in gamma ray output by the sources and the fact that the detectors are sensitive to gamma rays.

The neutron log is generally assumed to be against the side of the hole over the zones of interest. Figure 10-7 shows the results of the tool not being against the formation due to mudcake; the borehole cave being too short to allow the tool to enter or other such reasons. The log indicates too high a porosity if the tool is not against the side of the hole. Centered neutron logs, as were run in a few cases, made better calipers than porosity devices.

Note: Do not use gas bearing formations that are not well invaded as low porosity pivot points. These will throw your calibration off.

Figure 10-8 is a calibration of a neutron log using a logarithmic overlay given away by service companies. The 40% point was put at the shale line and 1% in a dense limestone or anhydrite streak.

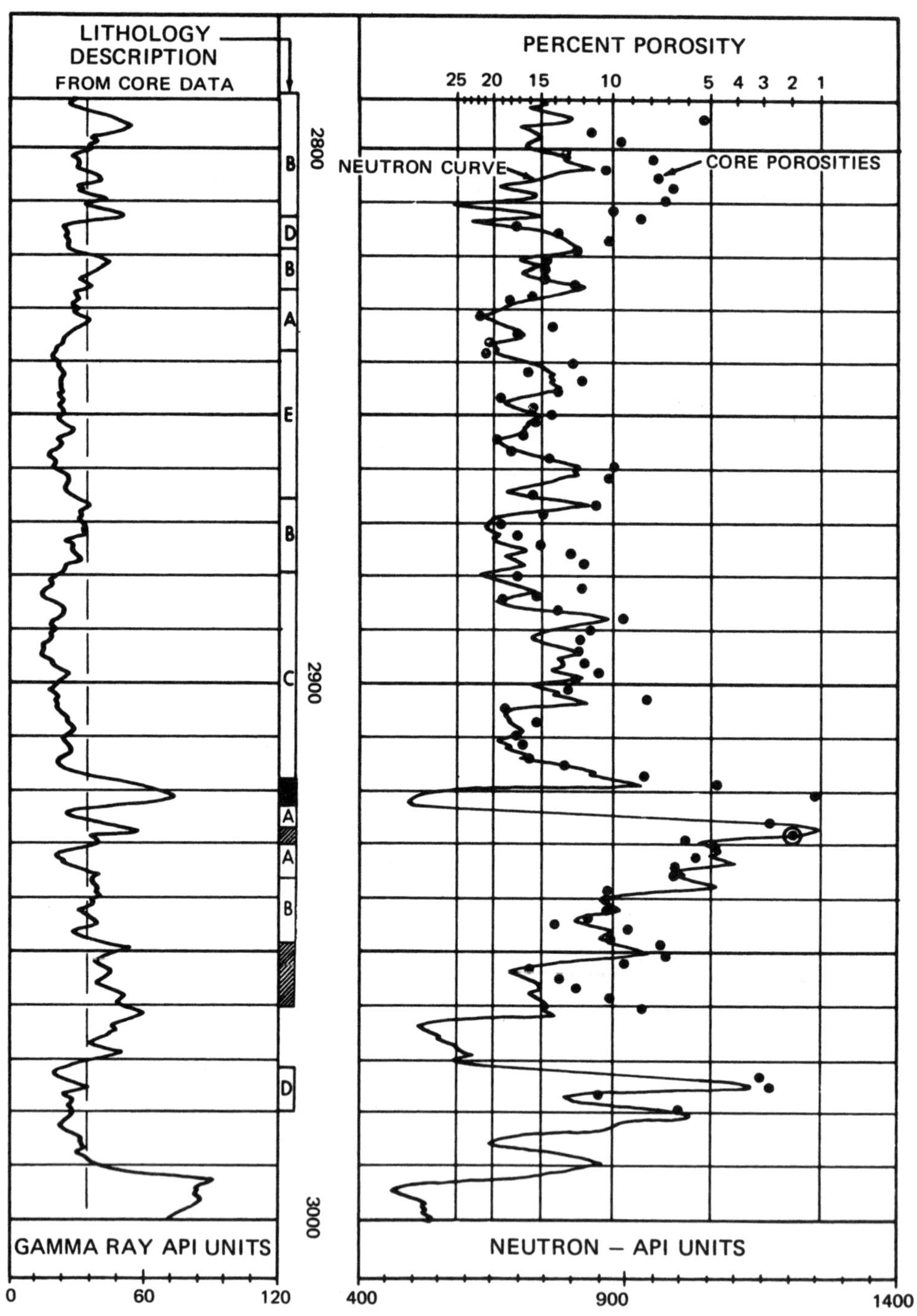

Figure 10-3. Neutron Deflection versus Core Porosity (courtesy Dresser Altas)

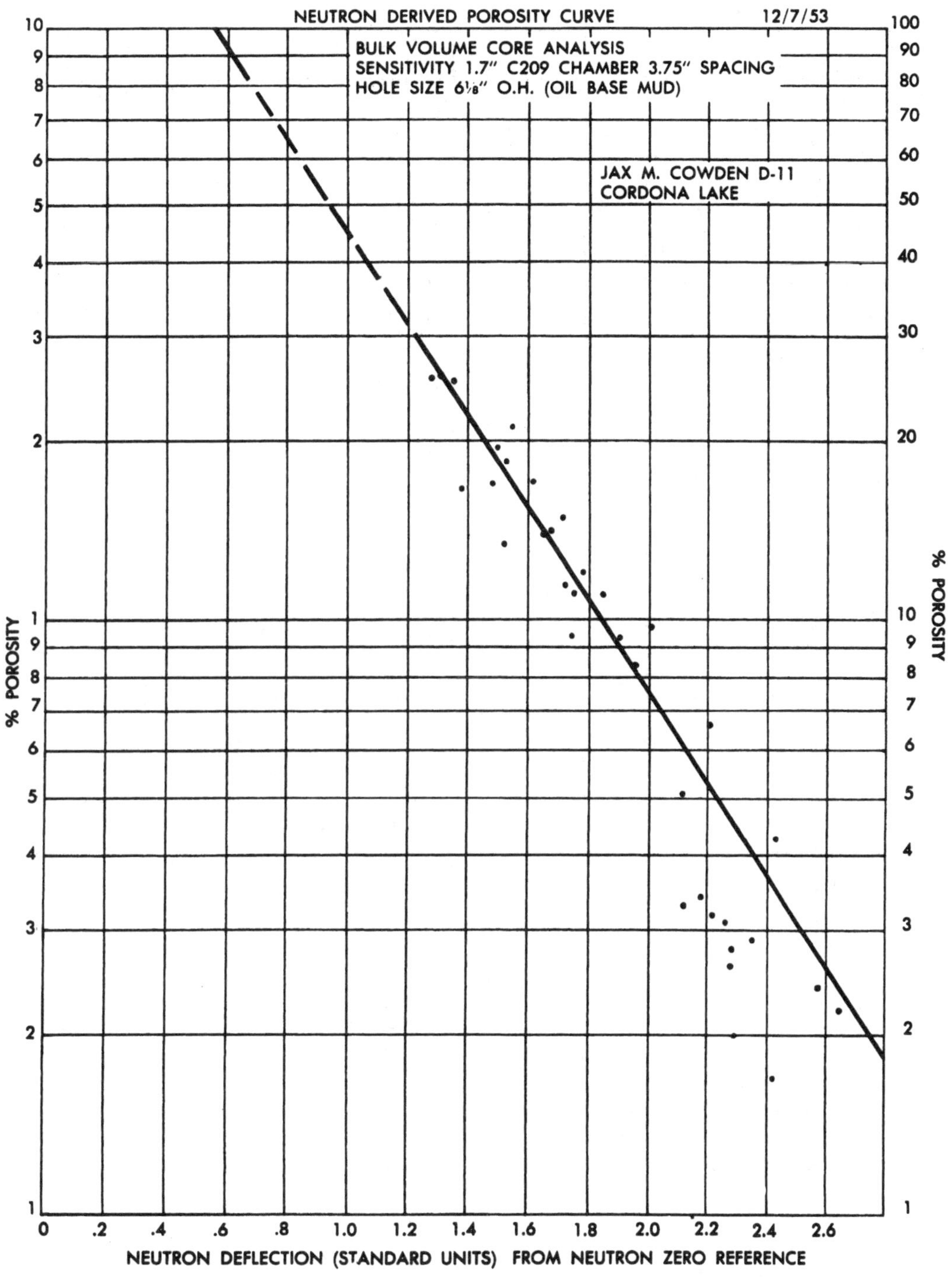

Figure 10-4. Calibration of a Neutron Log using Core Porosity

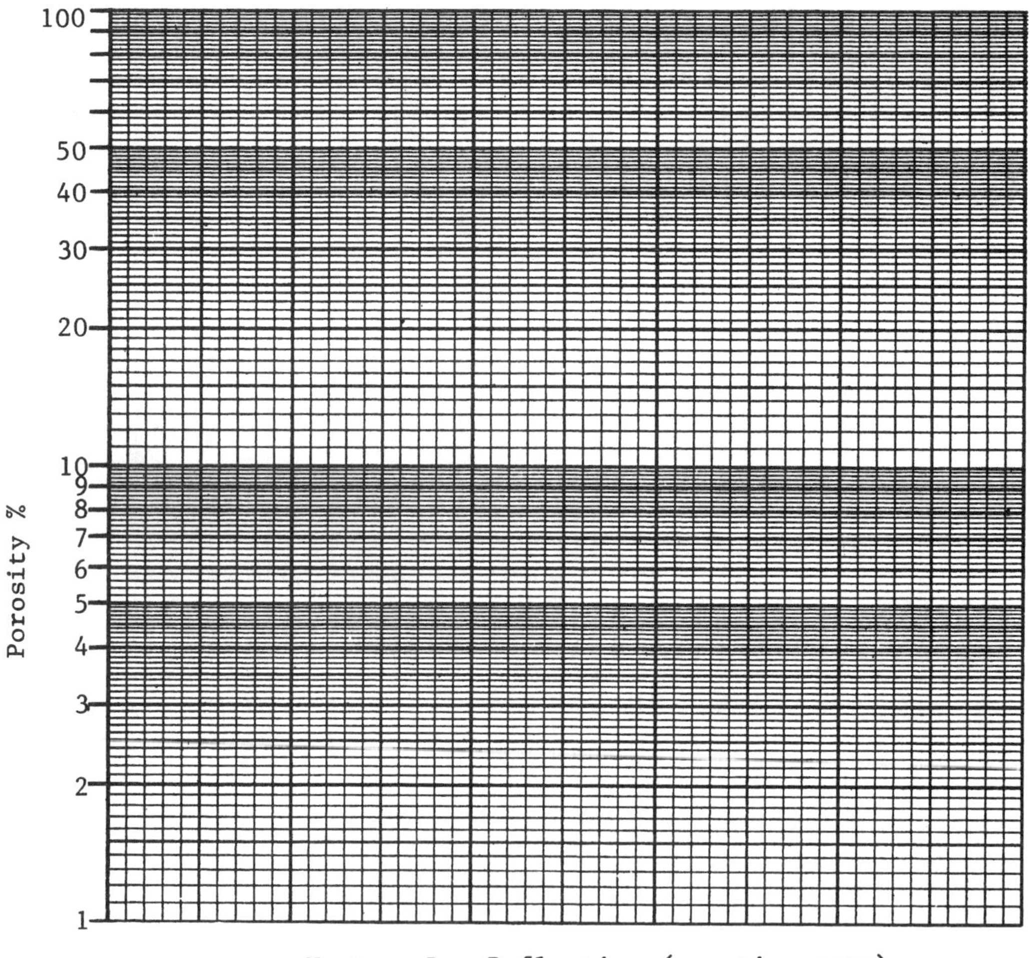

Figure 10-5. A Neutron Calibration Graph

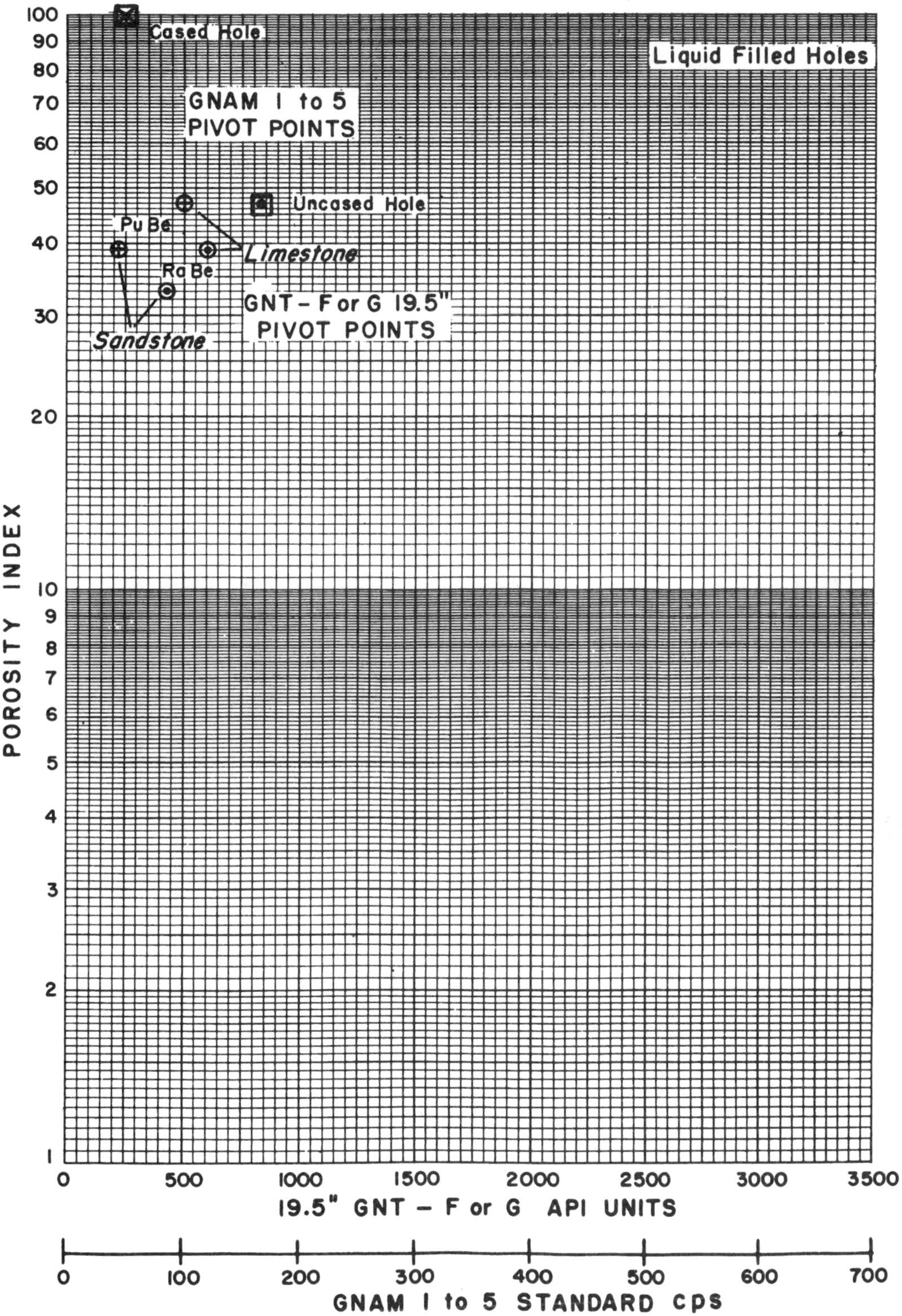

Figure 10-6. Schlumberger Pivot Plot Calibration Grid

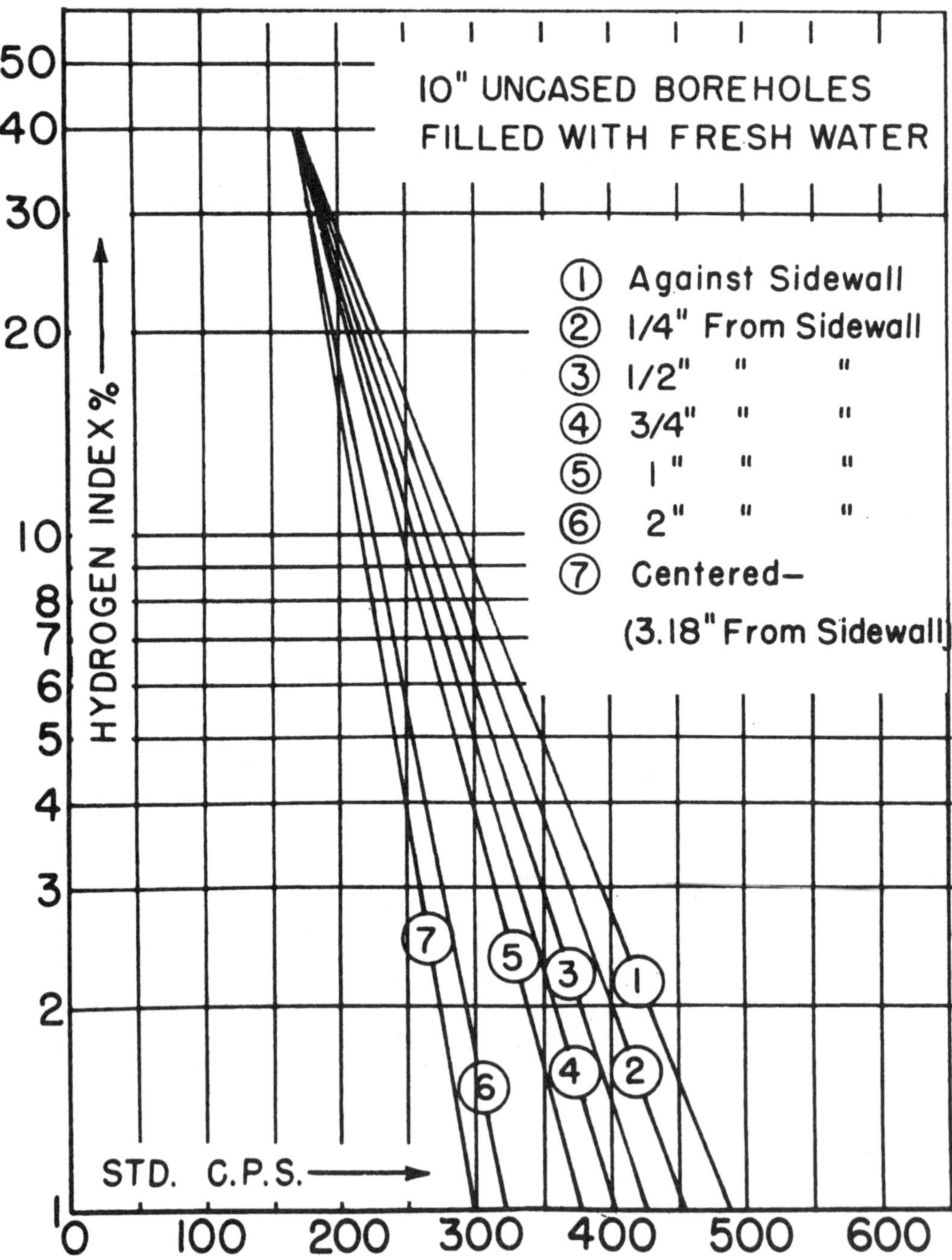

Figure 10-7. Effect of Distance of Sonde from Borehole Wall (after J. Tittman)

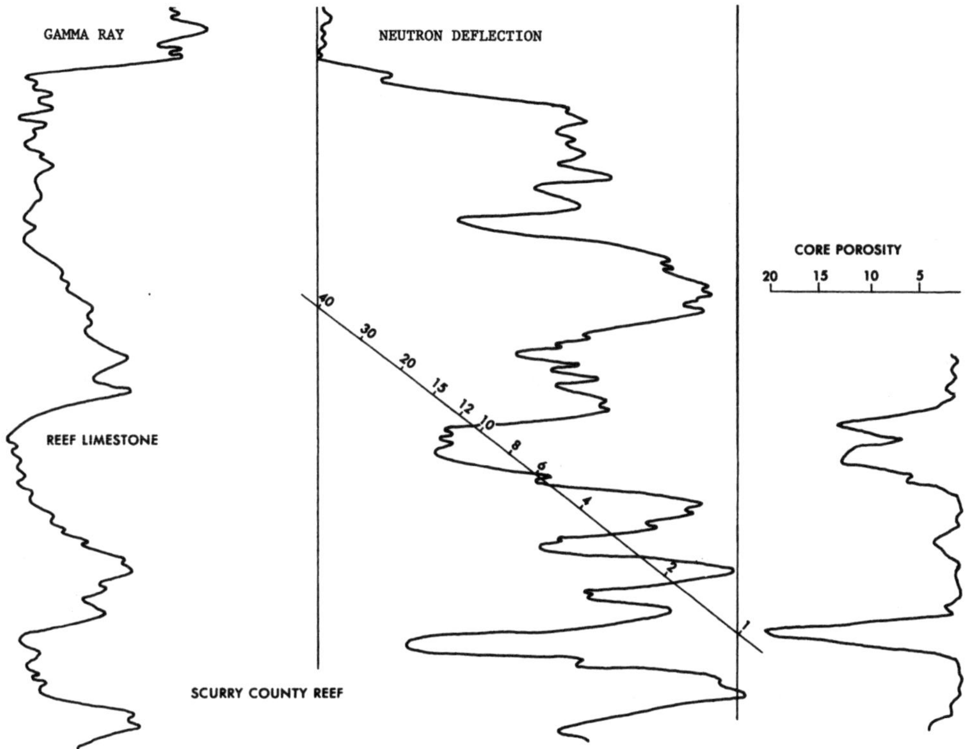

Figure 10-8. Neutron Log Calibration with Logarithmic Porosity Overlay

11

Hilchie, D. W., The electrical log and pulsed neutron capture logs, Old electrical log interpretation, p. 95–102.

The Electrical Log and Pulsed Neutron Capture Logs

Pulsed neutron capture logs such as the Neutron Lifetime and TDT logs are routinely run today in cased holes previously logged by electrical logs. These pulsed neutron capture logs (PNC) are devices that have much the same response as resistivity logs but are run through casing. They can be run any time during the life of the well to determine the change in fluid saturation in the reservoir. No effort is to be made here to teach you how to interpret the PNC logs except as a qualitative comparison with the ES.

The qualitative aspects of PNC logs are the same as with resistivity. Shales appear on the left side of the log, low porosity zones on the right side, and hydrocarbon zones on the right side. The parameter measured is "Capture Cross Section" with is designated sigma (Σ). There are usually several years between the time the ES was run and the later PNC. If nothing has changed with regard to saturation, the two logs should correlate well. Low porosity zones will show up as excursions to the right, hydrocarbon zones as excursions to the right and water zones as excursions to the left. If a zone was hydrocarbon bearing when the ES was run and was produced, resulting in water replacing the oil in the reservoir (which is common), the ES will look like a hydrocarbon zone and the PNC will look like a water zone.

The multitude of curves on some PNC logs will not be explained, only the sigma curve will be compared or discussed.

Figure 11-1 is a comparison of an ES and a PNC log in a well in the Texas Panhandle. Notice that the PNC sigma correlates very well with the short and long normals excepting in the 7710 – 25 zone which was an oil zone which was depleted by the time the PNC log was run.

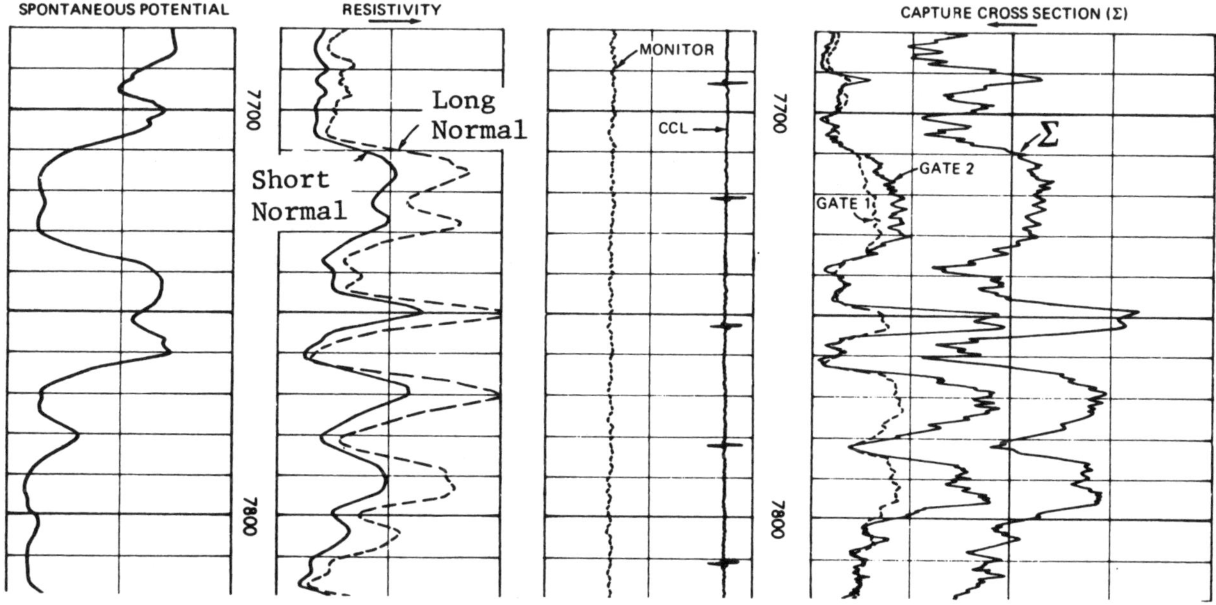

Figure 11-1. A Comparison of an ES and PNC Log

ELECTRIC LOG AND PULSED NEUTRON CAPTURE LOG COMPARISONS

Example 11-1 shows depleted oil zones in a well offshore Texas. Water has replaced the oil in the pore spaces. Low resistivity zone at 7530 – 50 is hard to see on the ES but is evident on the PNC log.

Example 11-2 is a 1945 ES with the lateral curve traced on the grid as it was run separately. A Neutron log not shown indicated the zones have essentially constant porosity of about 30%. Rw = .05. The ES shows oil zones at: A 4988 – 5008, B 5052 – 5068, and C 5093 – 5137 with water below. The PNC log shows zone C watered out. Note the high gamma ray

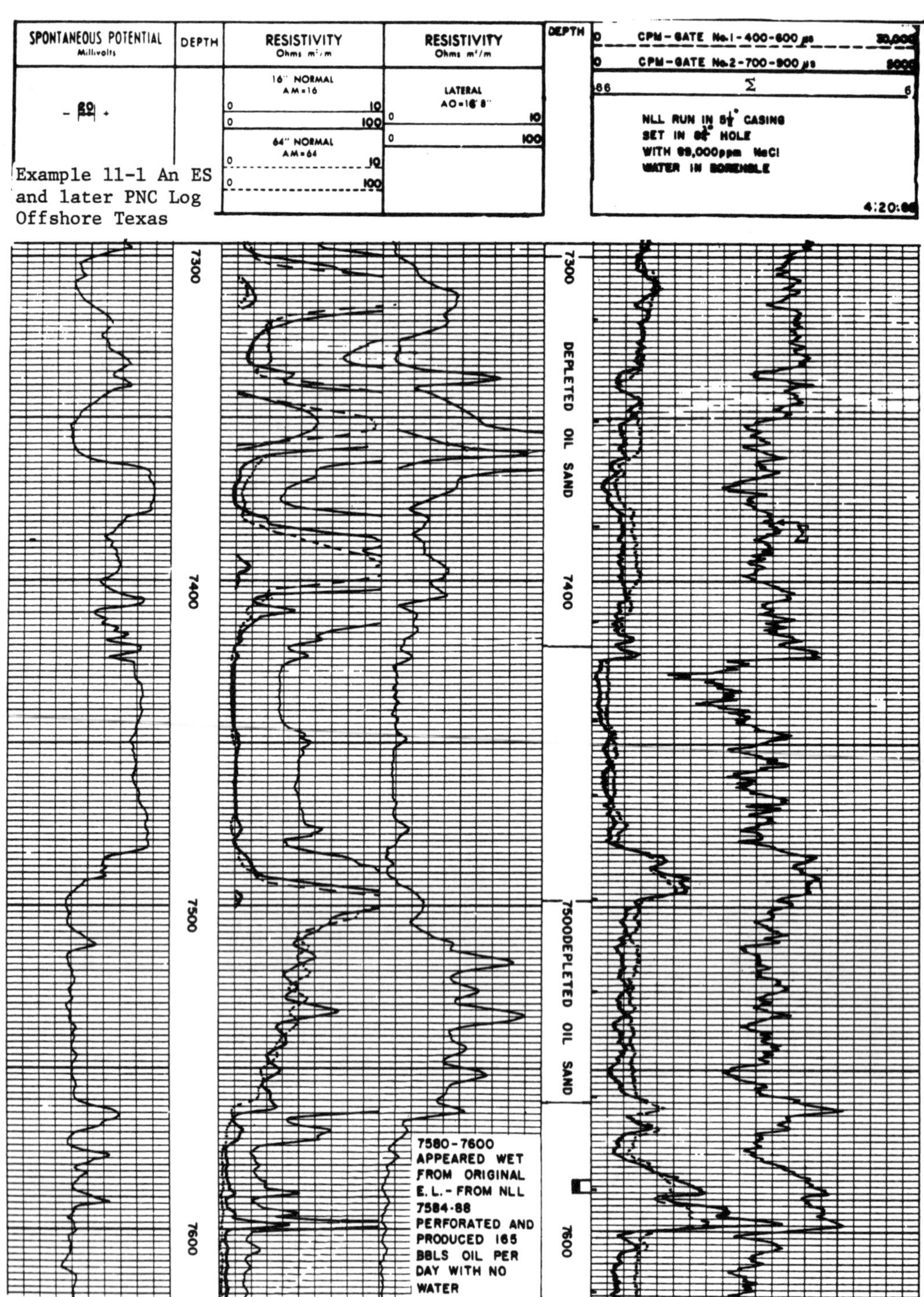

Example 11-1. An ES and later PNC Log Offshore Texas

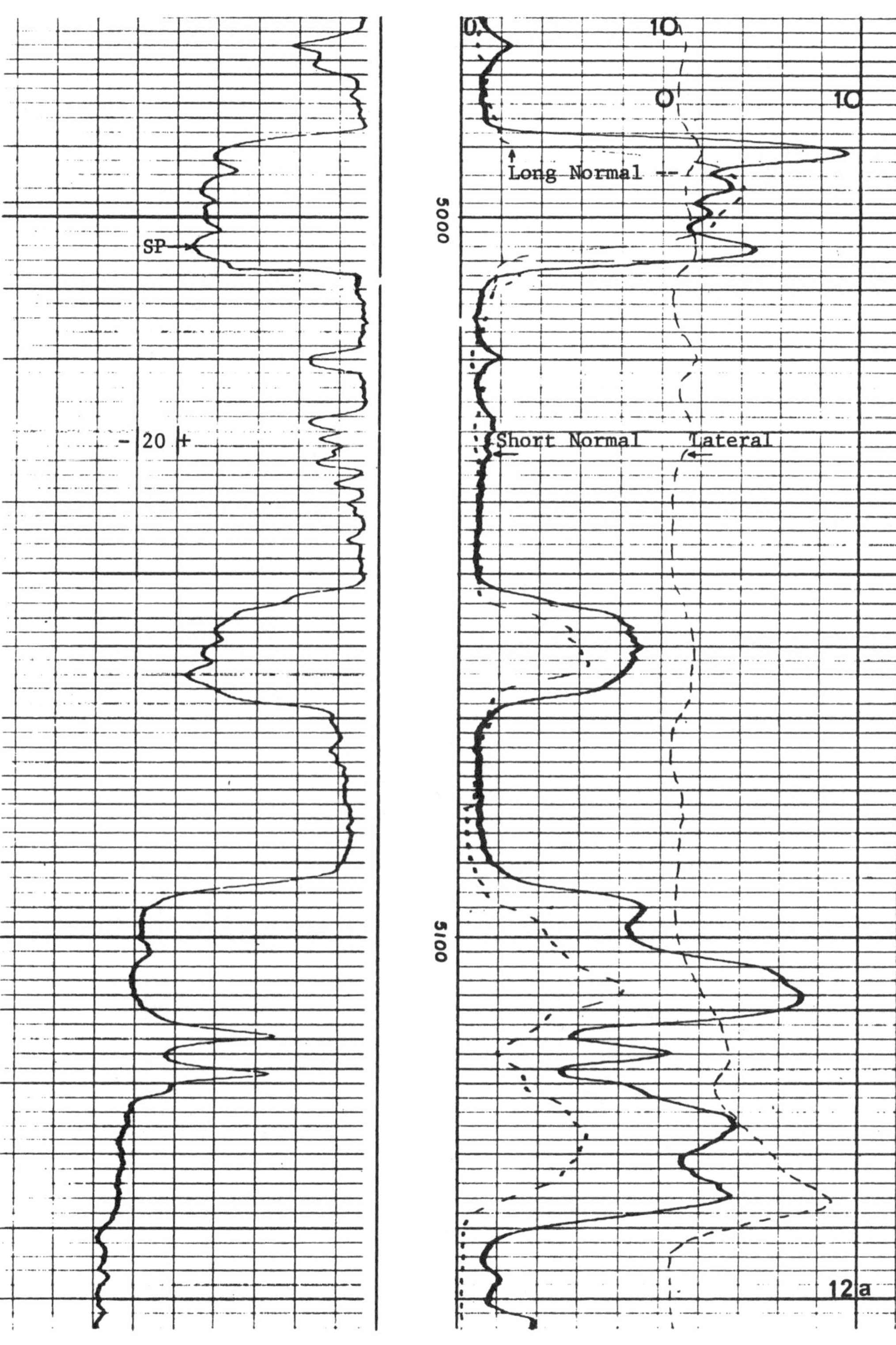

Example 11-2. South Louisiana 1945 Schlumberger ES - AM's 16″ & 63″ A0 24′. BHT = 130° at 5400 feet. Rm = 2.9 @ 86°

which results in this area when a well produces water and the radioactive salts plate out on the perforations and casing. Zone B is water on the PNC log. Zone A is still good.

Example 11-3 the PNC is drawn on top of the ES. Depleted gas zones are at 5690 – 95 and 5761 – 66. The PNC log still looks good, as water did not come in and replace the produced gas. The zone at 5657 – 71 is water and the zone at 5716 – 5721 is still possibly oil.

Example 11-4 is a 1959 ES with a PNC log traced on top. The lower permeable zone produced 75 BOPD

Example 11-2a. PNC Log for Example 11-2 Run Many Years Later

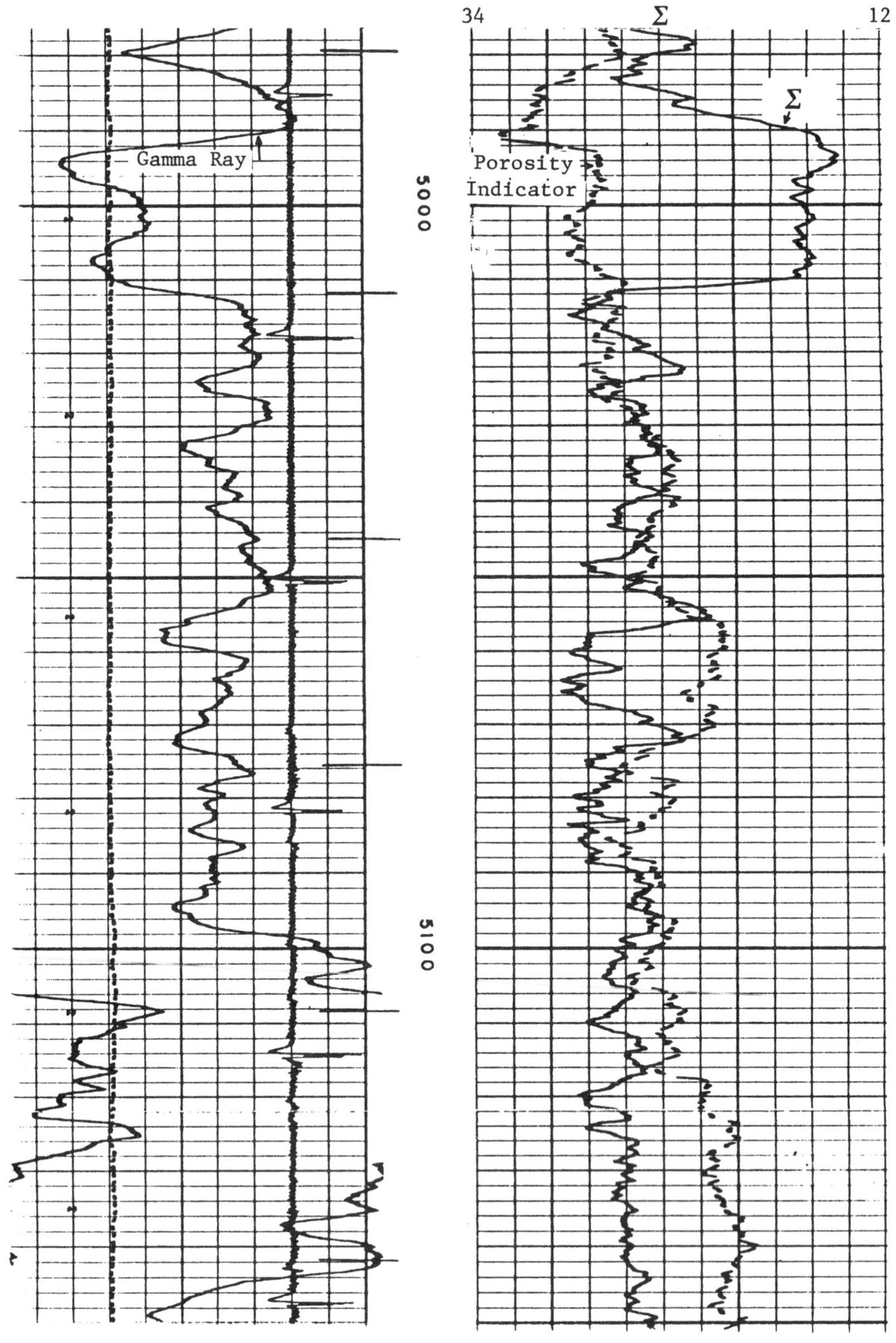

with a GOR of 20,000 from the perforations at 6034 – 37. There is obvious water below this zone. 5894 – 5930 is still possible oil. The rest appears to be wet.

Example 11-5 shows an old 1954 ES. The well was completed in the lower zone of 11385 –11433. Note the water below. The PNC log run later shows the bottom zone watered out. Upper gas zones are marked.

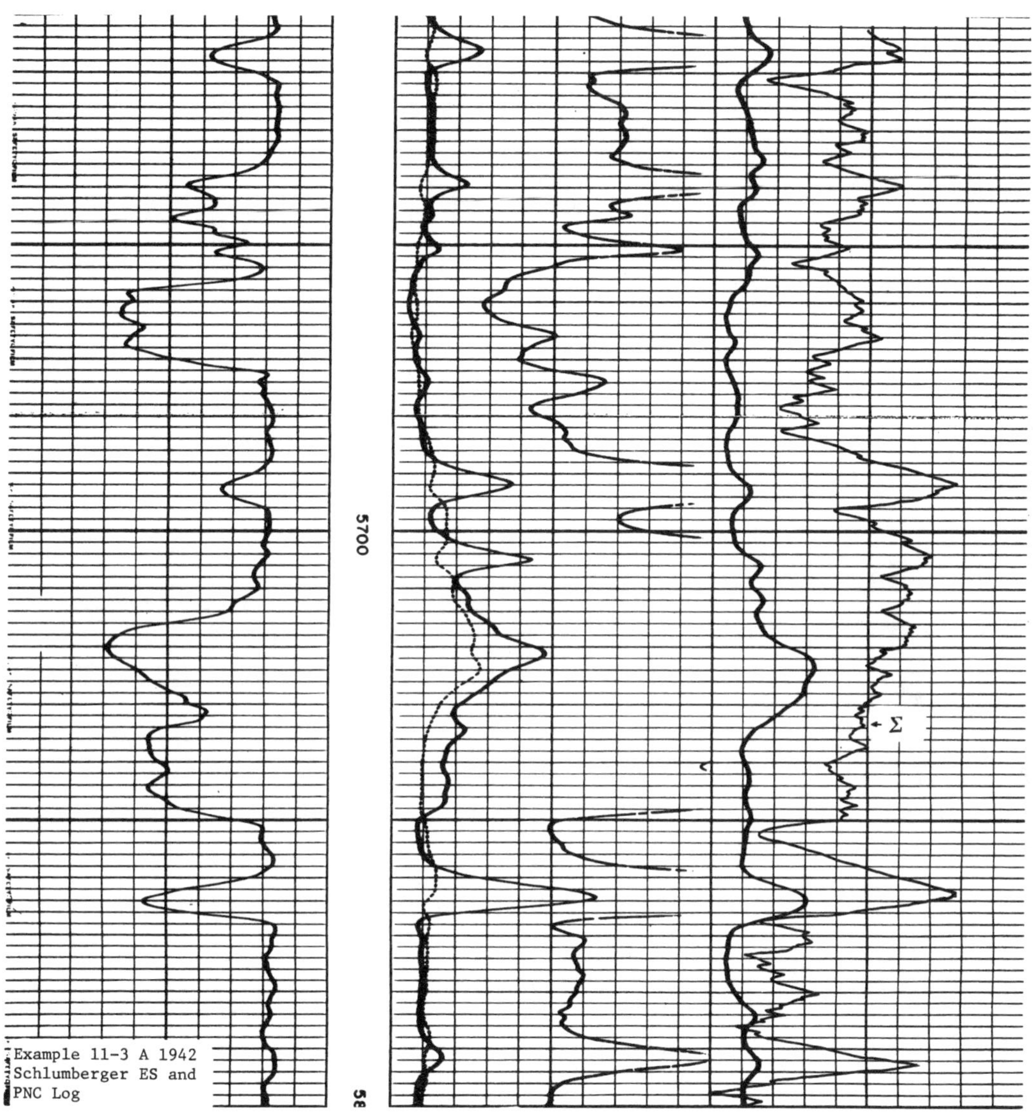

Example 11-3. A 1942 Schlumberger ES and PNC Log

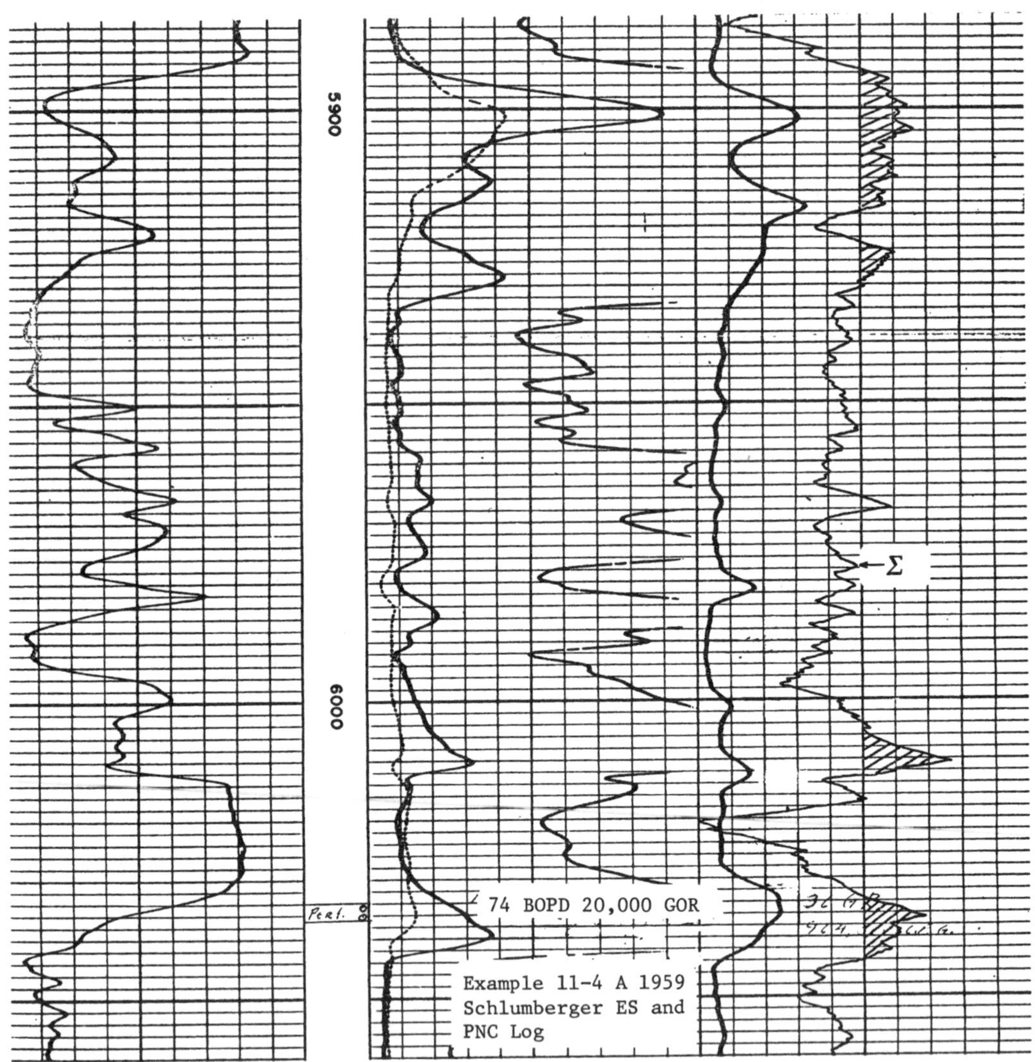

Example 11-4. A 1959 Schlumberger ES and PNC Log

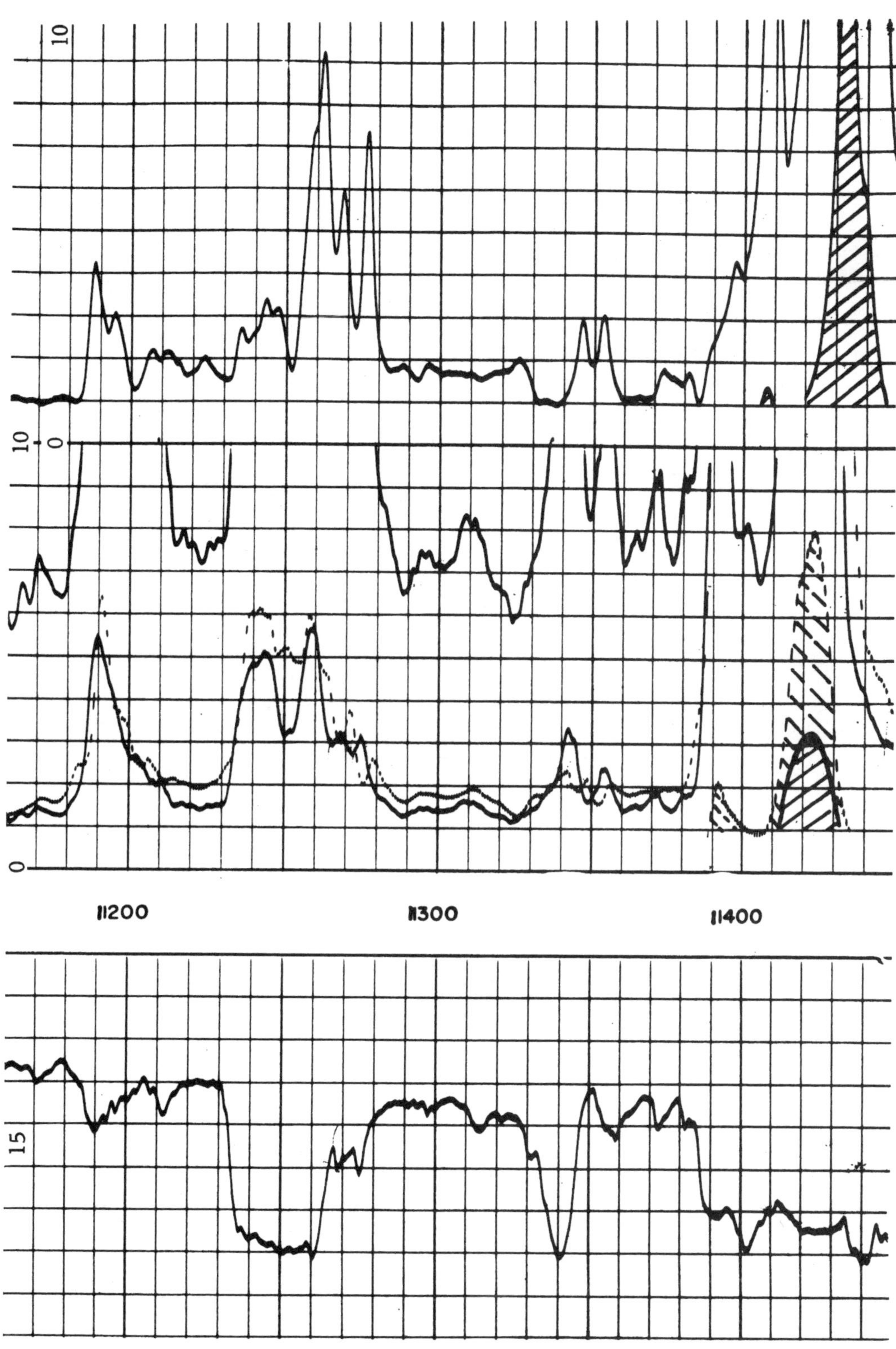

Example 11-5. 1954 ES — Gulf Coast

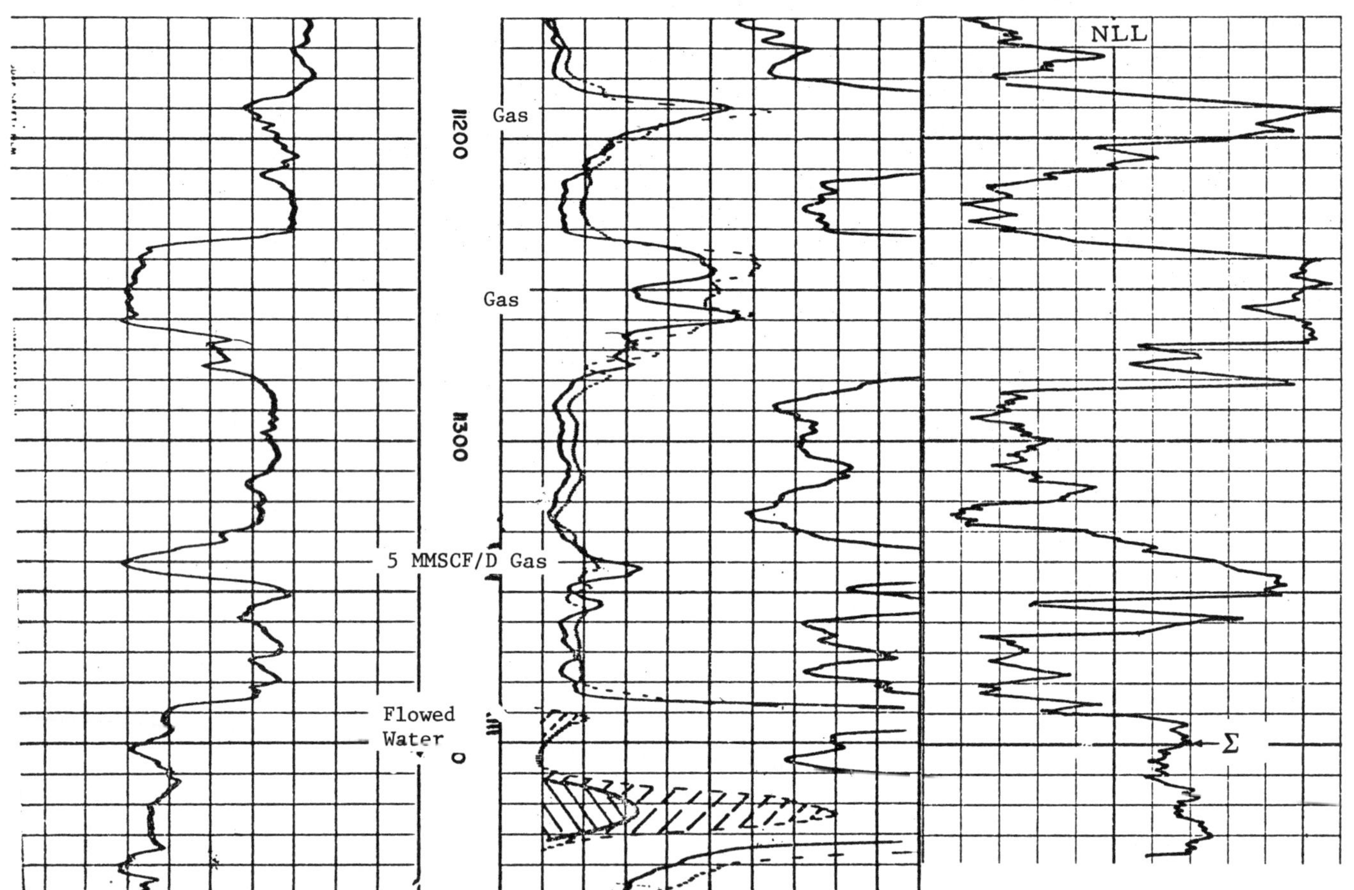

Example 11-5a. PNC Log Correlated to Part of ES from Example 11-5

Index

Symbols

A	acres (area) or current electrode
A0	lateral spacing
AM	normal spacing
a	constant relating porosity and F
B	return current electrode
d	borehole diameter
d_i	diameter of invasion
ES	electrical log
F	formation resistivity factor
Fa	apparent formation resistivity factor
I	current
L	length
M	voltage or potential electrode
m	porosity exponent of F
N	voltage or potential electrode (return)
n	saturation exponent (usually 2)
R	resistivity
Ra	apparent resistivity directly off curve
Rm	mud resistivity
Rmc	mud cake resistivity
Rmf	mud filtrate resistivity
Ro	formation resistivity with Sw = 100%
Rsd	resistivity of sand or formation that is permeable
Rsh	resistivity of shale
Rs	adjacent bed resistivity
Rt	true formation resistivity
Rw	formation water resistivity
Rwe	equivalent water resistivity
$R_{1\times1}$	microinverse resistivity
R_2	micronormal resistivity
r	resistance
T	temperature in degrees Fahrenheit
V	volume
V_{sh}	volume of shale
∅	porosity

Selected References

Schlumberger Resistivity Departure Curves (Document 1), May 1947
Review of Schlumberger Well Logging and Auximiary Methods (Document 2), July 1949
Resistivity Departure Curves (Document 3), 1949
Interpretation Hand-Book for Resistivity Logs (Document 4) July 1950
Interpretation Charts for MicroLog (Document 5), July 1952
Departure Curves for Laterolog (Document 6), August 1952
Resistivity Departure Curves (Beds of Infinite Thickness) (Document 7), 1955
Lateral Curves for Thin Non-Invaded Beds (a pamphlet), Sept. 1955
Introduction to Schlumberger Well Logging (Document 8), 1958

Guyod, Hubert Electrical Well Logging, a series of 16 articles in The Oil Weekly, Aug. 7 to Dec. 4, 1944
Guyod Model Electrical Resistivity Logs, 12 Volumes, 1954
Analysis Charts for the Determination of True Resistivity from Electric Logs (by H. Guyod and J. Pranglin), 1959
True resistivities from conventional electric logs Oil & Gas Jour. June 12, 1961

Chombart, L.G. Factors Involved in Practical Electric Log Analysis A Symposium on Subsurface Logging Techniques, Univ. Okla. 1949

Pirson, S.J. Formation Evaluation by Log Interpretation, World Oil, Apr. May & June, 1957

Lane Wells Log Interpretation Manual Number 2, 1956

Univ. of Kansas Fundamentals of Logging, Petroleum Engineering Conference Apr. 2 & 3, 1956

Dakhnov, V.N. Geophysical Well Logging, translated by G. V. Keller, Quarterly of Colorado School of Mines, Vol. 57, No. 2, Apr. 1962

Allaud, L. & Martin, M. Schlumberger, The History of a Technique, John Wiley and Sons, Inc., New York, 1977